"The ultimate professional's guide for understanding the ~~~~~ ~~~~~~~~~~ ~l
Backaler offers an extremely actionable, fast-paced, and enjoyable read."
—Aliza Licht, *fashion marketing executive and author of* Leave Your Mark

"To compete for consumers today, brands must create disruptive and shareable
experiences for the audiences that matter to them. In *Digital Influence*, Backaler
delivers an insightful guide on how you can align with the right influencers to execute meaningful, differentiated experiences that resonate."
—David Roman, *Chief Marketing Officer, Lenovo*

"Influencers are relevant for businesses in their home country as well as in the international markets where they operate. Backaler's book highlights the global implications of this trend, and what brands need to be aware of."
—Bruna Scognamiglio, *Vice-President Global Influencer Marketing, Gucci Beauty*

"The how-to book on influencer marketing the world has been waiting for!"
—Mark Schaefer, *author of* Return On Influence *and* KNOWN

"*Digital Influence* is highly actionable and well-researched – grounded in the first-
hand experiences of brand executives, agencies and influencers from around the
globe. Backaler's book is a must-read to understand how we got here, and more
importantly where we're headed."
—Nick Friese, *Chief Executive Officer, Digiday*

"Advocates/B2B influencers increase the number and reach of voices talking about
you. Beyond quantity, advocate content and interactions are more valuable because
buyers see them as authentic validation, not paid promotion. This is an incredibly
important topic for executives to understand."
—Laura Ramos, *Vice President and Principal Analyst, B2B Marketing,*
Forrester Research

"Influencers can help brands gain trust and authenticity with increasingly skeptical
and hard-to-reach customers. Joel Backaler provides in-depth guidance in *Digital
Influence* on how to achieve real results with influencer marketing globally. He gives
specific how-to tactical advice with rich examples and detailed case histories, all
based on his strong grasp of global marketing strategy."
—Charles Skuba, *Professor of International Marketing, Georgetown University*

Joel Backaler

Digital Influence

Unleash the Power of Influencer Marketing to Accelerate Your Global Business

Foreword by
Peter Shankman

Joel Backaler
Glendale, CA, USA

ISBN 978-3-319-78395-6   ISBN 978-3-319-78396-3   (eBook)
https://doi.org/10.1007/978-3-319-78396-3

Library of Congress Control Number: 2018939736

Cover illustration: elenabs/iStock/Getty

Printed on acid-free paper

This Palgrave Macmillan imprint is published by the registered company Springer International Publishing AG part of Springer Nature
The registered company address is: Gewerbestrasse 11, 6330 Cham, Switzerland

To my wife Qian:
You're my partner in all aspects of life—without your unfaltering support, love and encouragement, this would not have been possible.

To our new baby boy Henry:
It will be years before you understand how much you contributed to this book. Having your due date coincide with my manuscript deadline was all the motivation I needed to deliver on schedule.

Thank you both with all my heart.

Foreword

Go Google a YouTube video called "Supa Hot Fire Rap Battle." It's a parody of the classic "drop the mic" rap battles of the 1980s, featuring the star, our hero, who starts off every battle by announcing "I'm not a rapper," and then launches into, you guessed it, super hot fire raps, destroying his competition again and again. It's pretty funny.

I bring that up because meet anyone who considers themselves an influencer of any capacity, and the first thing they'll say to you is that they're not an influencer.

They'll tell you how much they hate that term. They'll explain that the term "influencer" is only used by people like Kim Kardashian, and OMFG, they're totally not like that.

Then, they'll start talking about their audience, and you'll hear lots of numbers, some very high, some very low, all with an explanation for why that specific number is the best number.

Then they'll throw out some fun terms, like "activation," or "conversations"—they'll definitely include "engagement"—and they'd hate themselves if they didn't use the phrase "the number of (pick one: retweets/likes) my client got was higher than anything they've ever had before" at least three times.

Then, on the flip side one day, you might find yourself at an industry conference of some sort, listening to someone who has an impressive title at a mid-size company giving a talk about how his company (under his direction, always under his direction) successfully ran multiple influencer campaigns that only cost (the total budget of your next six years' worth of marketing spend). Then, the speaker will show a PowerPoint slide that includes some media coverage of said influencer campaign and will smile, and smile, and smile, and somewhere else in the world, every time he smiles, a really cute animal is slaughtered.

Here's the thing, though, dramatic hyperbole aside: The one thing you'll almost never hear when it comes to influence and influencer marketing, whether from the influencer or the CMO talking about influence, is this—"how much money we made/product we sold/customers we acquired by doing what we did." That always seems to be a forgotten metric when it comes to any type of influence. And honestly? That's pretty sad.

You're about to dive into Joel's book for several reasons, but in the end, only two really matter: You want to find out what influencer marketing actually is, you want to see how you can use it for your business or company, and more importantly, you want to understand how it works, to learn how to stack the odds so that your investment in the world of social influence will generate a substantial return, time after time.

As I said, those are the only reasons that should matter.

Influencer marketing isn't new in the slightest, you know. Go back to high school. Remember the one popular kid who could come in wearing the most ridiculous outfit in the world (I'm thinking "Hypercolor" shirts), but because that kid was the cool kid, the next day, 30 people would show up to homeroom attempting to perfectly copy that look? That's influence.

Fast forward to your first car. Whether you know it or not, you were influenced to purchase it a minimum of 50 times, at least.

The only thing "new" about influence is the speed at which it now permeates everything we do, thanks to the fluidity of the social sphere in which we all live, work, and play.

Because of that new level of speed, influencer marketing has never been more important than it is right now, and it's only going to become more of a staple of every marketing plan.

But whether you use the knowledge in the following pages to your advantage, or it simply becomes another unused arrow in corporate marketing quiver that's unable to keep up with the times, will be determined by how you utilize your influencers, your audience, your products, and your brand as a whole, and how you keep them working in a symbiotic, harmonious orbit, that focuses on an overall goal of higher revenue, growth, and sales, while making sure to never overstep that magic, yet often hard to see line known as "ugh, they're trying too hard."

Joel Backaler is in the unique position to cast a shining light on the world of influencer marketing and illuminate for you what's working now, what will work in the future, and what will never work. Joel is an international marketing strategist, leading global marketing at Frontier Strategy Group, and is an award-winning blogger and Forbes columnist. He has lectured on this stuff around the world, and now he's ready to share it all with you. He's done this all in a down-to-earth, easy to understand style that will benefit any CMO or marketing decision maker at a company of any size.

And by the way: Why should you trust what I'm telling you? Well, I've started and sold three Internet companies in the past 15 years, I'm a corporate consultant to mega-brands around the world, including global hotel chains, airlines, restaurant conglomerates, and I'm on the major news networks talking about this stuff at least once a week.

But the most important reason you should believe what I'm saying goes back to what I said at the beginning of this foreword: I understand that having an audience is a privilege, and not a right, and I've spent the past 20 years of my professional career growing, nurturing, and caring for my audience by being real, honest, and down-to-earth. But hey, don't take my word for it. Just ask them.

Enjoy the book. You're going to learn a ton.

New York, USA

Peter Shankman
Best-Selling Author and Founder of
Help A Reporter Out (HARO)

Acknowledgements

To be able to write this book and dedicate the necessary time, research and attention to detail—while working a full-time job—required the tremendous support of a number of incredible individuals. I had well over a few hundred conversations and e-mail exchanges with the people listed below. I will be forever grateful for the perspective each of them shared with me over my time spent writing this book.

I would like to give a special thanks to the senior management team of Frontier Strategy Group, particularly Richard Leggett, Joel Whitaker, and Jon Rubin, for their flexibility and commitment. This work would not be nearly as strong without the thoughtful edits and feedback from Megan Close Zavala. The following individuals went above and beyond in their support: Mary Shea, Justin Szlasa, Dan Schawbel, EJ Lawrence, and Peter Shankman.

I wish I could detail how each person below made this book a reality, but the following individuals have my utmost gratitude and appreciation:

Soukaina Aboudou, Beca Alexander, Amber Armstrong, Evan Asano, Talia Baruch, Amy Backaler, Donna Backaler, Gary Backaler, Dez Blanchfield, Scott Brinker, Matt Britton, Shonali Burke, Reb Carlson, Tim Crawford, Ryan Crownholm, Veena Crownholm, Christian Damsen, Toby Daniels, Tom Doctoroff, Jeremy Epstein, Gil Eyal, Ziyang Fan, Kristy Fair, Jon Ferrera, Mark Fidelman, Nick Friese, Brendan Gahan, Amisha Gandhi, Chris Gee, Marie Han Silloway, Lauren Hallanan, Jim Harris, Yuping He, Brittany Hennessy, Alexander Hennessy, Madeleine Holbye, Carter Hostelley, Shel Israel, Walter Jennings, Cynthia Johnson, Carline Jorgensen, Mae Karwowski, Simon Kemp, Paul Kontonis, Michael Krigsman, Aliza

Licht, Daniel Liu, Jessica Liu, Laurent Magloire, Brigitte Majewski, Xavier Mantilla, Tamara McCleary, Andy Molinsky, Heidi Nazarudin, Ruben Ochoa, Lee Odden, Ryan Patel, Chris Purcell, Maddie Raedts, Delphine Reynaud, David Roman, Tuka Rossatti, Jill Rowley, Mark Schaefer, Bette Ann Schlossberg, Bruna Scognamiglio, Charles Skuba, Jerry Soer, Brian Solis, Taryn Southern, Josh Steimle, Victoria Taylor, Jai Thampi, Ronal Van Loon, Clark Vautier, Ashley Villa, Elijah Whaley, Kimberly Whitler, Tim Williams, Charles Windisch-Graetz, Heidi Yu, Julia Zhu, and Qin Zhu.

Contents

List of Figures

1

Introduction

Starting in 2011, with aggressive overseas acquisitions and offices opening around the world, Airbnb kicked off its international expansion. Fast-forward to 2018, and the company now operates in 15 countries and has enlisted more than 4 million[1] people to host strangers in their homes.

So many American companies have previously failed to translate their success overseas. There's a variety of reasons why brands fail to expand successfully, whether its failure to compete with local companies, an inability to tailor products for a new audience, or simply not doing the necessary up-front market research.

How did Airbnb do it? The right balance of global strategy and local implementation. Critical to success was the company's use of Local Influencers (individuals who can influence the actions/decisions of a loyal group of local online followers with regard to their particular area of expertise), to drive brand awareness through a series of cleverly designed campaigns. Some of the most memorable include:

"Local Lens Series" Paris
In early 2015, Airbnb launched an influencer content series featuring various thought leaders, creatives and experts in key cities sharing their views on must-see/must-try local experiences.[2] For example, in Paris, Airbnb worked with local bilingual food and travel writer Clotilde Dusoulier to author a "Local Lens" blog post on "10 Perfect Food Experiences in Paris," offering recommendations for "where to splurge on dreamy pastries" and "Paris' best baguette."

© The Author(s) 2018
J. Backaler, *Digital Influence*,
https://doi.org/10.1007/978-3-319-78396-3_1

"Night At" South Korea

In late 2015, as part of its "Night at" campaign and to support its expansion plans into South Korea, Airbnb worked with K-Pop sensation G-Dragon,[3] holding a contest where lucky winners could spend two nights in his recording studio in the heart of Seoul. This led to a flurry of social media and traditional media attention, as well as a spike in Korean Airbnb user registrations.

"Don't Go There, Live There" London

In 2016, Airbnb's London team designed an exclusive experience townhouse in London, where visitors could come and experience local food and music. The purpose was to show travelers "the real London" not found in guidebooks. It enlisted 25 Local Influencers to amplify the event on social media—while 1400 guests visited the townhouse over four days, several millions learned about it online as a result of the influencer engagement.[4]

These Local Influencers—both celebrities and online personalities—opened the door for Airbnb in market after market across the globe. Why should Airbnb directly tell customers how great their service is? Better to leave it to Local Influencers who can speak with authenticity and authority.

Brian Chesky, co-founder and CEO of Airbnb, explains, "By early 2011, we were primarily an American company. But it became very clear that international is really important. We're a travel company. Us not being international is like your phone not having email… So it became very clear that we had to be international – we had to be a GLOBAL travel network."[5]

For Airbnb, Local Influencers made a foreign brand a local one. That's the difference.

***

My day job is as an international marketing strategist at Frontier Strategy Group, where I help senior executives from many of the world's largest brands. Part of my job is to pay attention to major global trends, and every now and then I am captivated by something I am seeing in the marketplace and decide to write a book like this one. I dive deep, obsess over a million details, and attempt to synthesize what I have learned into actionable, executive-level findings.

If I have done my job, I will keep readers like you a step (or two) ahead of the next big thing. That was the point of my last book, *China Goes West*, which tells the story of Chinese brands expanding beyond China's borders to take on Western multinationals.

When I set out to write *Digital Influence*, I intended to tell a story that was primarily an international one, like the experience of Airbnb that begins this introduction. I thought the story would be about brands that found international growth by using Local Influencers to jump-start their expansion.

When I started research, however, I discovered the international side of influencer marketing is only a small part of a much bigger story that demanded to be told. And when I started interviewing "influencer marketing insiders" like marketing executives, agencies, and influencers, things started to get messy…

Influencer marketing is young and unsettled. I found I needed to address many fundamental questions before I could consider introducing the international story.

What types of questions you may ask?

Take the most basic term, "influencer." It is highly disputed within the industry—in Peter's foreword, for example, he reveals that not even influencers like to actually be given that name. Instead, alternative terms get thrown around like "content creator," "talent," "KOL," "YouTuber," "blogger"… if you can read Chinese, there is even "网红".

Beyond a lack of standard terminology, there is also a lack of industry best practices across each phase of working with influencers:

- How do you identify the right influencers for your brand?
- What does it take to get an influencer to respond positively to your outreach?
- Which forms of brand–influencer collaborations are most effective?
- How do you measure Return On Investment (ROI) and also make sure you can trust the data you're using for measurement?

One of the major points I make in the book is influencer marketing is a global phenomenon that is developing rapidly, and companies can tap into Local Influencers to advance their international strategy. But that is far from the only aspect of what you are about to learn.

In fact, you are about to get an in-depth look into a topic that is generally only written about at a surface level, with clickbait headlines about "How Many Thousands of Dollars Brand X Paid Influencer Y for a Post" or "The Top 10 Steps to Getting Instagrammers to Promote Your Brand."

I am going to take you from the basics, to the practice, to the global relevance. And ultimately, I will give you a peek at where we are headed.

You are not just learning from me. You will hear from more than 100 individuals I interviewed during the writing of this book, ranging from Fortune 500 executives to high-growth start-up founders to agency leads to influencer software vendors to influencers from around the world.

There is a lot packed in here. I hope you find the insights valuable— both today when you think about how these case studies, frameworks, and methods can be applied to your business, and later on, when you use it as an ongoing resource to build (or rebuild) your company's approach to influencer marketing.

A few final words before we begin our journey together:

Definitions for terms like "influencer" and "influencer marketing" vary from person to person—to make sure we are all on the same page, this book is intentionally written to explain concepts and define key terms early on that continue to be used in that manner throughout the book.

This book covers a wide range of industries from fashion to enterprise software to insurance—keep an open mind about how examples from one industry could be applied to your business.

Influencers will only become more important to the way we engage our target audiences in the years to come. The days of "interruption marketing" through disruptive ads are ending—people want to learn from trusted peers, not faceless companies. Now is the time to bring some much-needed consistency and standardization to the practice. I aim to open your eyes to the broader global implications of how influencers can contribute to global growth and share how your organization can unleash the potential of influencer marketing to accelerate your business.

Let's begin.

Notes

1. Hartmans, Avery. "Airbnb Now Has More Listings Worldwide Than the Top Five Hotel Brands Combined." *Business Insider*, 10 Aug. 2017, www.businessinsider.com/airbnb-total-worldwide-listings-2017-8.
2. Dusoulier, Clotilde. "Local Lens: 10 Perfect Food Experiences in Paris." *The Airbnb Blog Belong Anywhere*, 25 Feb. 2015, blog.atairbnb.com/local-lens-10-perfect-food-experiences-in-paris/.
3. Airbnb. "Night At G-Dragon's Second Home." *Night At—Airbnb*, Oct. 2015, nightat.withairbnb.com/case_studies/gdragon.html.
4. Hiorns, Benjamin. "Airbnb Launch the 'Live There' House in Shoreditch with a #DingDong." *Creativepool*, 29 June 2016, creativepool.com/magazine/advertising/airbnb-launch-the-live-there-house-in-shoreditch-with-a-dingdong.9811.
5. OneSky Content Team. *How to Ace Global Marketing Like Airbnb*, pp. 9–9, http://offers.oneskyapp.com/hubfs/OneSky%20Airbnb%20Global%20Marketing%20Slideshare.pdf.

2

Then vs. Now: Influencer Marketing (Re-)Defined

Susan starts the morning like she does most Mondays: The jarring sound of the alarm on her iPhone interrupts her final moments of peaceful slumber. Tired from a weekend spent celebrating her 26th birthday, she picks up her phone, switches off the third and final snooze alarm, and begins to review various news, social media, and messenger applications.

Meanwhile, Robert, a 23-year-old sales associate at a tech start-up, is trying to make up for a weekend spent consuming excessive amounts of delivery pizza and energy drinks while watching eSports and playing video games with friends. He works up a sweat using a seven-minute full-body workout app and then refuels with a protein shake recipe he discovered on YouTube. The countdown clock is ticking, but Robert still has a few minutes left before he needs to leave for the office, so he quickly skims over his personal e-mail inbox and checks out a few gaming websites.

At face value, Susan and Robert appear to have on their smartphones unlimited, free access to all the content they could ever want to begin their weeks: "What's the weather?" "Who won last night's game?" "When will my package arrive?"—a simple text or voice search delivers immediate access to their desired information.

Susan and Robert are fully aware online content isn't free—it comes at a cost in the form of their time and attention. While they access that information on their smartphones, Susan and Robert are constantly annoyed by disruptive ads.

© The Author(s) 2018
J. Backaler, *Digital Influence*,
https://doi.org/10.1007/978-3-319-78396-3_2

Here are just a few "payments" they each make during their early morning smartphone sessions:

- **Checking e-mail**:
  - Spam offer for 15% off a revolutionary new mattress Robert doesn't need

- **Scrolling Facebook**:
  - Feed advertisement for a sustainably sourced wooden laptop case Susan would never buy

- **Watching YouTube**:
  - A heart attack-inducing pre-roll movie trailer for an upcoming horror movie that streams before Robert can view a protein shake how-to video tutorial

- **Viewing News Websites**:
  - Ever since she shopped for a friend's baby shower gift, a series of ads for a "Snoogle" over-sized pregnancy pillow intrusively follow Susan from website to website

Over the course of their morning smartphone time, disruptive advertising tactics like these invade seemingly every piece of content they try to access. The sad truth for marketers seeking to reach Susan and Robert with these techniques is they're wasting their money—and worse—they're damaging their brand.

For example, if a marketer surveyed Susan later that morning and asked:

What brands do you remember seeing a moment ago on your Facebook feed? Susan would likely respond, "None."

If she could recall a brand name, it would be because she was particularly frustrated or annoyed with the advertisement—one that left her asking, "How did they know I like this? Are they reading my emails?" She might decide it's time to finally download the ad-blocking software her friend told her about. It's not just Susan. People are becoming less and less trusting of brands.

Unfortunately for Susan and Robert, these types of invasive ads continue all day long. Susan, however, does get a break from them on her commute to work. She's developed a daily morning ritual to help pass time on her 40-minute subway ride into the office. When Susan finally gets on the subway, she puts in her earphones, closes her eyes, takes a deep breath, and presses play on her favorite healthy-living podcast—she never misses an episode.

After a brief introduction, the podcast host pauses for 30 seconds to talk about a show sponsor, a company that has just come out with a new, cost-effective healthy meal kit service designed especially for busy millennial professionals like Susan. She listens to the host with deep interest as he talks about his experience cooking a delicious meal the other night and how much he loved the food and the process.

Before transitioning back to the episode, the host shares a special discount code for a free meal. Susan figures, "What's the risk? It sounds delicious and podcast host X hasn't let me down yet."

Just before the train arrives at her office station, Susan redeems the special discount code and places her first meal kit order. Her mouth waters at the thought of a healthy, home-cooked dinner of garlic kale, steamed fragrant quinoa, and sesame-glazed chicken breast—prepared in just 15 minutes or less! (Fig. 2.1).

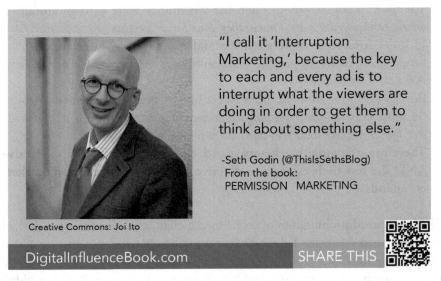

"I call it 'Interruption Marketing,' because the key to each and every ad is to interrupt what the viewers are doing in order to get them to think about something else."

-Seth Godin (@ThisIsSethsBlog)
From the book:
PERMISSION MARKETING

Creative Commons: Joi Ito

DigitalInfluenceBook.com   SHARE THIS

Fig. 2.1 Seth Godin

Interruption Marketing Is Dead. Long Live Influencer Marketing!

What did the healthy meal kit company do differently? How did it succeed in grabbing Susan's attention where the mattress, laptop case, and Snoogle companies all failed? Seth Godin, considered by many industry insiders as the "godfather of modern-day marketing," may have the answer.

In his best-selling book, *Permission Marketing*, Godin refers to the traditional approach that most marketers rely on to gain consumer attention as "interruption marketing.[1]" He argues that people are busy and nobody is eagerly waiting to read the latest magazine ad or watch a new commercial. However, marketers rely on *interrupting* day-to-day experiences in hopes that their efforts—no matter how disruptive—will plant a seed in our subconscious to take action (make a purchase) at a later date.

He describes a vicious cycle where marketers spend more to reach consumers and receive less in return[2]:

1. Human beings have a finite amount of attention so they can't watch everything—*if Robert has 20 minutes to watch YouTube, he must choose between learning a new protein shake recipe or watching video game reviews*
2. Human beings have a finite amount of money so they can't buy everything—*if Susan only has $300 of discretionary spending each month she needs to carefully decide where to spend it*
3. The more products brands offer, the less money there is to go around—*once a consumer buys an iPhone, they don't buy a Samsung smartphone*
4. In order to capture more attention and more money, interruption marketers must increase spending—*however, this increase in marketing exposure costs a lot of money*
5. But, by spending more in order to get bigger returns, consumers encounter more clutter—*their viewing experience is increasingly filled with disruptive brand messages that they don't want to watch*
6. **Therefore, the more marketers spend, the less their efforts are effective <u>AND</u> the less their efforts are effective, the more interruption marketers spend.**

The widespread proliferation of social media channels and connected devices in recent years has provided interruption marketers with even more ways to reach consumers, but Godin's vicious cycle remains constant—humans only have so much time, attention, and money to spend. As a result, Susan didn't want to pay attention to unwanted e-mails, Facebook feed ads, or YouTube pre-roll ads. She was too busy concentrating on e-mails from her colleagues, Facebook status updates from her friends and family, and reading the latest news headlines.

Unlike the other marketing messages that infiltrated Susan's hectic morning, the healthy meal kit company did something differently. It collaborated with a trusted figure in Susan's life (the podcast host) to deliver a marketing message to Susan indirectly through a relatable story about cooking a healthy dinner.

We've all heard podcasts that include either interruption marketing messages or influencer-led permission marketing messages. When podcasts use interruption marketing messages, the tone completely changes. Imagine if the meal kit company used a typical interruption marketing approach to sponsor the podcast. When Susan pressed play, instead of hearing the familiar voice of her favorite podcast host, she would hear something like this:

Deep, generic radio voice:

Our meal kits are the best! We use healthy, organic ingredients and meal preparation is SO convenient – especially for busy millennial professionals. Go to our website today and sign up to take advantage of a special offer for new customers.

Result: Susan ignores the ad, just like all the other ones she heard earlier that morning.

So, what happens when Susan learns about the meal kit company from the podcast host instead of a generic radio voice?

First, she opens her ears. The podcast host is a trusted, familiar voice, a voice she regularly listens to for advice on how to improve the quality of her life.

Second, she listens. Susan relates to the podcast host's personal story about cooking his own meal and is pleased to hear about the positive results.

Third, she takes action. After receiving a *recommendation* (note: not interpreted as an intrusive ad) from a trusted voice, and receiving an attractive offer from the company for a free meal, Susan sees no downside in trying the meal kit service.

This is how influencer marketing differs from interruption marketing. Susan's story also helps illustrate that while the Internet and technology have enabled a new generation of influencer marketing to exist, the fundamental drivers for what makes influencer marketing effective lie in word of mouth recommendations from trusted peers. *Today, word of mouth recommendations*

"We've always known that word of mouth is the most impactful way to communicate. For brands this means we need to create an interesting story that other people will talk about."

-David Roman (@iamdavidroman)
Chief Marketing Officer | LENOVO

DigitalInfluenceBook.com   SHARE THIS

Fig. 2.2   David Roman

can happen in a "one-to-many" fashion via virtual relationships, like Susan's bond with the podcast host she never met; but in the past, word of mouth's ability to spread was much more limited (Fig. 2.2).

Word of Mouth: The Ultimate Enabler of Influence

One of the most famous word of mouth "marketing campaigns" in US history occurred on April 18, 1775, when a silversmith named Paul Revere learned of an oncoming British attack—a march that would begin the next day and result in the arrest of prominent colonial leaders and the seizure of the local militia's guns and ammunition. Revere knew he needed to get the word out, while there was still time to prepare local colonial leaders for the British offensive.

In his book *The Tipping Point*, Malcolm Gladwell explains what happened later that evening when Revere began his "midnight ride" to warn leaders in different towns along the way.

> [Revere] knocked on the doors and spread the word, telling local colonial leaders of the oncoming British, and telling them to spread the word to others. Church bells started ringing. Drums started beating. The news spread like

a virus as those informed by Paul Revere sent out riders of their own, until alarms were going off throughout the entire region.[3]

In the end, Revere's word of mouth marketing campaign was a success. By the time the British began their march, the colonial militia was ready to fight back, and their triumph eventually led to the American Revolution.

Word of mouth's effectiveness hasn't changed much since the time of Paul Revere and the American Revolution. The influential recommendations of trusted peers not only cause us to take action, but often compel us to share that recommendation with others who could benefit in our communities. However, "Community" is no longer limited to the people we know in the real world, and messages no longer rely on traditional channels and gate-keepers to reach their target recipients. Today, everyone has the ability to influence large audiences, and it's largely due to the advance of new technology and the marginalization of industry gatekeepers, who once had ultimate authority to choose who could be influential over the masses.

Chris Gee, managing director and head of digital at global public relations firm Finsbury, explains, "As we get more inundated with marketing messages, the trust in those types of messages tends to go down and we look more to the opinions of people who are like us. When you take all the technology away from it, it makes total sense. If I'm going to look for a restaurant for dinner, I'd much rather get a recommendation from a trusted friend over watching 50 ads to go to that same restaurant."

There are even industry leaders who argue word of mouth is the basis for all marketing efforts:

"The role of marketing as a discipline doesn't change much over time," said David Roman, chief marketing officer of Lenovo. "We've always known that word of mouth is the most impactful way to communicate. For brands this means we need to create an interesting story that other people will talk about – that has and will always be at the center of everything we as marketers do."

But, it takes a special type of individual to trigger the word of mouth chain reaction. Not just anyone has the authority or charisma to spread ideas virally across communities of peers. This special type of person is what Gladwell refers to as *mavens*—information specialists who expose us to new concepts and ideas. They play a unique role in setting trends and shaping new ideas. Mavens are not only experts in their chosen domain, they naturally feel compelled to share that knowledge with others who may benefit. Gladwell sums up the role they play in society as "information brokers, sharing and trading what they know."[4]

The healthy-living podcast host is an example of a modern-day maven in his niche (unique information about healthy-living) for his specific audience (people like Susan who want to live a healthy lifestyle and consider his advice on the topic to be authoritative). He feels a deep sense of responsibility to constantly seek out the latest information and trends to help his listeners. By extension, his audience respects the host's expertise and wants to learn from him (similar to the way colonial leaders respected Revere and his important message). His recommendations have so much influence on their lives that audience members not only actively listen to each podcast, but also feel compelled to share relevant recommendations with friends in their real and online worlds who could benefit.

In this respect, mavens are the ultimate influencers. They impart their expertise on individuals who trust their opinion and then those individuals listen and take action based upon the maven's recommendation. This aligns well with the Word of Mouth Marketing Association's definition of influence: *"the ability to cause or contribute to another person taking action or changing opinion/behavior."*[5]

After cooking her own delicious dinner, Susan will surely share her experience testing the meal kit company with other friends and colleagues who are also trying to eat healthy, thus fueling a series of word of mouth recommendations across her community.

Historically, "community" was restricted only to people the influencer directly interacts with—those tightknit social circles limited by social, cultural, and geographic barriers.

For example, let's say that while Susan may listen to podcasts in her spare time, Robert reads car review magazines. Previously, Robert could only share his knowledge of the latest engine advancements with a small community: his immediate family and maybe a few friends from work. This is great news for Robert's inner circle, because whenever someone needs a new car, they know exactly who to ask. But, only a handful of immediate connections can tap into Robert's expertise for the same advice.

What if Robert had a bigger platform to share his automotive expertise with a much larger audience? How many more people could he potentially influence to make the right car purchase decision?

On the other extreme, Paul Revere had to physically ride a horse from town to town to share his message about the impending British threat. Revere had critical, time-sensitive information that all colonists wanted to know, but the message had to slowly pass from person to person as Revere gradually reached each town. Imagine how much easier it would have been if Revere had the ability to send a quick tweet to a targeted list of colonial leaders with the hashtag #TheBritishAreComing (Fig. 2.3).

paulrevere1775 @paulrevere1775 Apr 18
Watch out for the British!!! #TheBritishAreComing #MidnightRide

DigitalInfluenceBook.com   SHARE THIS

Fig. 2.3 Paul Revere "Tweet"

Rising Technologies, Disappearing Gatekeepers

> TV is resilient, but the only thing that's really resilient is the Super Bowl – that means the Super Bowl is a museum to a previous era. It's fun to go visit once a year, but those days are gone. What matters these days are community, word of mouth and real value. Influencers are a key part of that equation. Social media lit the fire on this – influence is much more democratized.
>
> —*Jeremy Epstein, CEO—Never Stop Marketing*

Epstein, the former chief marketing officer of the billion-dollar social media software firm Sprinklr, explains how the media landscape has shifted. Thirty years ago, traditional media such as television, radio, newspaper, and magazines were the only game in town.

On the one hand, during traditional media's golden age, consumers had significantly less choice about what they could watch, listen to, or read compared to available options today. Consumers willingly offered advertisers their precious time and attention in exchange for the latest news, information, and entertainment. On the other hand, it was really challenging for everyday individuals to influence people outside their local communities and immediate social circles (e.g., Robert, the automotive expert from the previous section).

Would-be influencers faced three major obstacles to build their audience and become known under the traditional media system:

1. Content was very expensive for an individual to produce and promote—*there were no iPhones to film videos or take digital photos*—*expensive high-end production equipment was unattainable by everyday individuals*
2. Creating content required specialized knowledge and expertise—*there was no YouTube to learn from tutorials on the fly or online marketplaces to cost-effectively hire freelancers*
3. Promoting content required approval from industry gatekeepers—*there was no way for an individual to reach a large audience through their own efforts.*

Because traditional media prevented everyday individuals from gaining influence, celebrities became the ultimate influencers. Movie stars and high-profile TV personalities could influence large audiences with their widespread star appeal. They were the ones brands hired to help shape the opinions of mainstream audiences and were used as part of a seemingly authoritarian, top-down advertising approach—think about how celebrities featured in fashion magazines used to dictate "what's hot" in each season.

The widespread adoption of connected devices like smartphones and use of myriad social media platforms have broken down traditional barriers and enabled a new generation of influencers to share their message with expansive audiences of Internet-connected online citizens.

"If I wanted to be known 25 years ago, I'd have to be in the newspaper and on TV programs very frequently and there would have to be some gatekeeper [a TV producer or news editor]. Someone else would have to make a decision for me to be known," explains Mark Schaefer, social media strategist and author of the book *Return on Influence*.

Shaefer drives home the point that today everybody can publish content, share their expertise, and build an audience:

> We don't have to go through gatekeepers and wait to be picked – we can pick ourselves.

Lilly Singh unknowingly applied Shaefer's remarks to her own life when she started a YouTube channel in October 2010. She had just dismantled a small dance company in Toronto that she spent years trying to make work. Her teammates let her down time and time again with their lack of commitment; they would show up late or miss practices, while Singh stayed up throughout the night producing marketing materials to promote the team's next show. She was sick of relying on others who didn't possess her same ambition and drive.

That all changed when she discovered YouTube. The shift from disgruntled dance troupe leader to independent YouTube content creator was completely liberating for Singh. Reflecting on this period, she wrote:

> I remember feeling a new sensation the first time I uploaded a video. I wrote the script, shot it, edited it, and released it. No one else was involved or required, and the independence was exhilarating.[6]

Singh poured her creativity and entertainment experience into engaging content that helped her build a loyal community of 12 million + YouTube subscribers. Her satirical videos cover a range of topics from everyday life, including arguing with parents, hip hop culture, young adult struggles, and much more. For added comedic effect, she often dresses up to play outrageous versions of characters like her own parents and the cast of the HBO series, *Game of Thrones*.

For all intents and purposes, Singh has become a celebrity. At the time of writing this book, her most recent videos feature mainstream celebrities such as Hollywood actor Dwayne "The Rock" Johnson and pop star Selena Gomez. She has even collaborated with established international brands like Coca-Cola and Toyota. She's living the celebrity lifestyle as well: At age 26, she moved from Canada to Los Angeles, where she purchased a $1.5-million Spanish-style house and a brand-new Tesla to drive around town[7]—all with her hard-earned YouTube cash.

Singh's experience is certainly an extreme example. It's increasingly difficult for individuals to replicate her massive success as countless copycats flock to YouTube and other social media platforms with hopes to achieve similar results. However, Singh's story does illustrate what is possible today in a world with fewer gatekeepers and multiple avenues for individuals to build an audience entirely through their own efforts. Think about it: Singh built a global audience of more than 12 million subscribers without receiving permission from anyone but herself. Her own YouTube channel was the vehicle that enabled it all to happen.

Which transitions to the final point of this chapter: As much as the ubiquitous adoption of social media has enabled a wider range of individuals to access large audiences, the other side of the story is the resources required to create engaging content are more accessible today than ever before.

Specifically:

1. *Technology has advanced, making the equipment required to produce high-quality content like video much more cost-effective and easier to use for everyday people*
2. *Online connections through freelancer marketplaces enable individuals to connect to specialized talent (ex: graphic designers or video editors) on-demand that is both cost-effective and capable of delivering professional results.*

What once required thousands of dollars, a large production team of specialized talent and expensive equipment can now be achieved by an individual in a home studio with a smartphone camera and an external microphone.

Through online talent marketplaces, individuals can efficiently and cost-effectively hire skilled graphic designers, video editors, voiceover talent, website developers, and a wide range of other professionals to help with the production process. Many of these specialists live in markets around the world where hourly rates may be significantly lower to produce a high-quality final product.

The combined result is that individuals have access to broader audiences via social media and they can leverage more accessible technology and talent resources to produce high-quality, impactful content to fuel audience growth and drive engagement over time.

Redefining the Relevance of Influencers in Our Modern World

At this point, it should be clear that we're not talking about something brand-new and revolutionary. Influencer marketing and the power of word of mouth have a long-established track-record of causing people to take action. However, there are a series of factors that make influencer marketing more relevant today than in the past.

First, everyday consumers have a growing distrust for brands and an aversion to traditional interruption marketing techniques that infringe upon their access to free content—so much so that they are increasingly downloading ad-blocking software.

Second, in a world with so much content and so many channels, many brands are stuck in Seth Godin's vicious cycle, one where they spend more and get less in return. While traditional advertising remains effective, it's imperative for the modern-day marketer to concentrate on engaging communities and building positive word of mouth—influencers are a major part of this equation.

Third, influencers were once limited to mainstream celebrities, but as social media reach audiences, and the resources to create high-quality content became more available, a new generation of influencers has emerged. While mainstream celebrities still have a role to play in influencer marketing, there are new ways for different types of influencers to work with brands. What is relatively new and what has led to all the "buzz" around influencer marketing is technology's role in enabling a fast-growing number of everyday people to have a voice and build an audience by showcasing their passions and expertise, and in turn giving a platform to potential influencers who otherwise would never have been discovered before the widespread adoption of social media.

So, who are these influencers? And what do marketers need to know when considering collaborating with them?

That's exactly what you are about to learn in Chapter 3: *Levels of Influence: Key Characteristics of Modern-Day Influencers.*

Marketer's Cheat Sheet

- **Interruption Marketing vs. Influencer Marketing**: We are transitioning from an era of interruption marketing (where advertisers intentionally disrupt people's activities to plant a branded message into their subconscious) to one where community and word of mouth play a much more significant role (especially influencer marketing).
- **Consumers More Receptive to Influencers**: Susan views the podcast host's story of the meal kit company as a recommendation, not as an ad. Given her existing virtual relationship with the host, she opens her ears, listens, and takes action. This is in stark contrast to interruption marketing messages that consumers either ignore or suppress with ad-blocking software.
- **Word of Mouth 2.0**: While word of mouth has been around since early man gave the first recommendation to a trusted peer, what has changed significantly is the role word of mouth plays in a world where individuals are connected to more like-minded peers than ever before, and internet-enabled word of mouth spreads messages and builds influence at never-before-experienced speed.
- **Celebrities as the Original Influencers**: Mainstream celebrities were the original influencers, and they still have an important role to play, but the door is now wide open for a much more diverse set of individuals with engaged audiences to share the spotlight and collaborate with brands.
- **Decline of Gatekeepers**: Individuals previously required permission from gatekeepers to access large audiences (think: TV producers, news editors, etc.), but today through social media including blogs, YouTube, Facebook,

Twitter, LinkedIn, Instagram, WeChat, etc., an individual can build a large, active audience through his or her own efforts. Gatekeepers still exist, but they no longer pose as significant of a barrier as they once did.

- **Advance of the Creator's Toolkit**: Professional-looking content—videos, images, audio recordings, white papers, webinars, and so on—previously out of reach for everyday people, is now more accessible than ever before. Today it's easier than ever for individuals to use affordable production tools and "rent" specialized talent on demand allowing anyone to produce high-quality, impactful content to fuel audience growth and drive engagement over time.

Notes

1. Godin, Seth. *Permission Marketing: Turning Strangers into Friends and Friends into Customers.* New York: Simon & Schuster, 1999, 25–27. Print.
2. Godin, Seth. *Permission Marketing.* New York: Simon & Schuster, 1999, 38. Print.
3. Gladwell, Malcolm. *The Tipping Point: How Little Things Can Make a Big Difference.* New York: Hachette Book Group, 2000, 30–31. Print.
4. Gladwell, Malcolm. *The Tipping Point.* New York: Hachette Book Group, 2000, 69. Print.
5. Word of Mouth Marketing Association. *The WOMMA Guide to Influencer Marketing.* 2017, pp. 7–7.
6. Singh, Lilly. *How to Be A Bawse: A Guide to Conquering Life.* New York: Ballantine Books, 2017, 6–7. Print.
7. Landau, Emily. "Lilly Singh Goes to Hollywood." *Toronto Life*, 23 Mar. 2017, torontolife.com/city/life/inside-dizzying-world-lilly-singh-torontos-accidental-megastar/.

3

Levels of Influence: Key Characteristics of Modern-Day Influencers

It's a cloudy fall day in San Francisco back in 2001, and Tim Ferriss, an entry-level sales associate, stares blankly at his computer screen in silent disbelief. He has just found out that after a year of grueling 12-hour workdays, he is the second-lowest-paid employee in the entire 150-employee data storage company, and his earning potential will not improve anytime soon. "This isn't what was supposed to happen," he thinks.

Rewind to a little over a year before: Tim graduates from Princeton University with a degree in East Asian Studies and flies from New Jersey to California to start his career. News of a friend successfully selling a business for $450 million inspires Tim to journey west in search of his own Silicon Valley riches. Three months after graduation, he still doesn't have any solid job leads, despite it being a hot job market. So, in an act of brute force, Tim sends 32 consecutive emails to the CEO of the data storage company, then a 15-person start-up. Naturally, the CEO figures anyone with this level of persistence could become a great sales professional, and that's what Tim is hired to do. As the company grows, however, it becomes more and more evident that it is a dead-end role. Efforts to work with his manager to improve sales processes are quickly shot down with blanket justifications such as, "Because I say so."

Back to fall 2001: As Tim stares at his computer, he realizes that he can't achieve financial independence by relying on an incompetent manager or support from fellow corporate drones. He decides to take matters into his own hands: He stops working and starts using his workdays to do online

© The Author(s) 2018
J. Backaler, *Digital Influence*,
https://doi.org/10.1007/978-3-319-78396-3_3

research. Instead of cold calling the next prospect, he learns how to out-source everything from manufacturing to ad design. Within a few months, he uses this know-how to launch an online nutritional supplements busi-ness that practically runs itself. With his newfound freedom, Tim travels the world and learns new skills including Argentine tango, Japanese horseback archery, and Chinese kickboxing.[1]

The experience of designing a new life—both professional and personal—changed Tim's outlook on what success looks like. While he ultimately didn't replicate his friend's $450 million payday, he did achieve a different type of independence: He no longer made $40,000 a year at the data storage com-pany, but he could earn $40,000 per month with his online company. He didn't need to wait for retirement to travel the world; he did so while still working on his own business. He knew he was onto something that could be replicated, and more importantly, he felt compelled to share his experi-ence with others, initially in the form of his best-selling book, *The 4-Hour Workweek*, first published in 2007.

For more than 10 years, Tim has evangelized the concept of "lifestyle design," which he discovered through his own self-transformation. In the process, he has accumulated an extremely loyal and engaged community which voraciously consumes his content, debates its findings, and shares it across the Web. His deep connection with these community mem-bers, and his understanding of their likes and dislikes, has enabled him to engage them to help launch four more best-selling books and a top-ranked podcast with more than 200 million downloads. Most impor-tantly, as Tim's profile grows, he remains authentic to his followers, only creating content and promoting services that he knows they will value and never sacrificing his audience's trust in exchange for a big one-time payment.

By all measures, Tim is a modern-day influencer. He has a loyal online following of fans who interact with his content, take his advice, and buy products he recommends. Tim is a different type of influencer from Lilly Singh, the YouTuber introduced in Chapter 2, however, both in terms of the type of content he produces and who he produces that content for. And certainly, both differ greatly from Robert the car enthusiast, who only influ-ences his immediate friends and family. However, all three exert their influ-ence on their unique audiences in different ways via the channels they have access to.

If they are all so different, then how can all three individuals be classified under the same umbrella classifier "influencer"?

There are a few ways of slicing and dicing influencers, but the best way is probably through a series of different lenses based on the size of the community they influence, the platforms they use, and the role they play for companies that engage them.

Levels of Influence: Celebrity Influencers, Category Influencers, and Micro-Influencers

"Raise your hand if you're an influencer." Carter Hostelley, chief executive officer of Leadtail, a B2B social marketing agency, addresses a packed room of senior executives from an American Fortune 500 technology company.

Nobody in the audience raises their hand in response.

Well, there are two people with raised hands—Hostelley and Jill Rowley, a social-selling expert. They are among a handful of speakers who have been invited to present during an all-day digital strategy workshop.

"Let me share a definition that might make a few more of you raise your hands," Hostelley says as he goes on to read the BusinessDictionary.com definition of "influencers" for the group:

> Individuals who have the power to affect purchase decisions of others because of their (real or perceived) authority, knowledge, position, or relationship.[2]

Finally, hands start to go up. By the time Hostelley finishes sharing the definition, about 60% of the attendees have raised hands.

That's much better than where they started, but for the roughly 40% of the audience who still have their hands down, Hostelley has another question for them: "How many of you have ever convinced a friend or loved one to try a new restaurant or buy a new product?"

This time, *everyone's* hand goes up.

Herein lies, at the most basic level, one of the most confusing aspects of influencer marketing. The term "influencer" is used so broadly, and for so many different types of people in so many different situations, that nobody envisions the same "influencer" when the topic comes up in conversation. Each person will naturally have their own perceptions of what an influencer is based on their individual context and background.

For example, let's look at what might have been going on in the heads of three of Hostelley's workshop participants when he first said, "*Raise your hand if you're an influencer*":

I'm no Kim Kardashian. I don't even have an Instagram account.
—*Alex, senior vice president of sales enablement*

I haven't written a long-form essay since I graduated from college. I couldn't even imagine becoming a blogger.
—*Julia, senior director of search marketing*

Who does he think I am, Leonardo DiCaprio? Nobody cares about what I post online.
—*David, director of strategy*

It wasn't until Hostelley clarified the broader definition of "influencer" that attendees gradually understood how they could each be influencers for at least a small, niche audience.

But how helpful of answer is it to say, "everyone's an influencer," especially when marketers want to understand which influencers best fit their brand? Chapter 7 provides more specific guidance for brands to identify the right influencers for their goals, but at a high level, it's easier to first break influencers down into three general buckets based on the size of audience[3] they have influence over. The three levels of influencers below are all (i) *known for something* among a target community and (ii) can *influence the actions* of target-community members.

The three groupings are Celebrity Influencers, Category Influencers, and Micro-Influencers. To see how these three levels of influence apply to a single category takes a moment to review the following image. The following are illustrative examples of three influencers for the broad category of "automotive" (Fig. 3.1).

1. **Celebrity Influencers**: Celebrity Influencers possess broad-based fame and are able to influence a mainstream group of fans. Traditional celebrities like actors, athletes, and musicians could be considered Celebrity Influencers, but this category also includes online personalities with massive followings—either Category Influencers (see #2 below) who have achieved celebrity status or other high-profile individuals who have worked their way through the traditional gatekeeper system to achieve equivalent recognition and influence.
   On the flip side, someone who's just famous for famous' sake, but does not have the expertise or ability to influence a target community, is simply a celebrity, not a Celebrity Influencer.

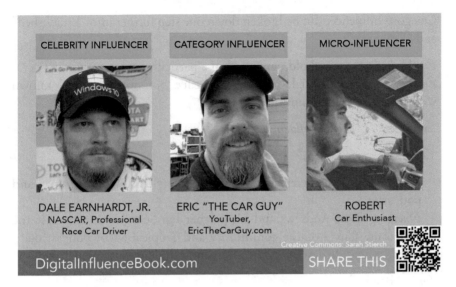

CELEBRITY INFLUENCER	CATEGORY INFLUENCER	MICRO-INFLUENCER
DALE EARNHARDT, JR. NASCAR, Professional Race Car Driver	ERIC "THE CAR GUY" YouTuber, EricTheCarGuy.com	ROBERT Car Enthusiast

Creative Commons: Sarah Stierch

DigitalInfluenceBook.com   SHARE THIS

Fig. 3.1 Examples of Celebrity, Category and Micro-Influencers

Examples of Celebrity Influencers:

- Jessica Alba, CEO of The Honest Company
  –Actress turned executive, influential over buying decisions of mothers seeking products that are both safe and effective for their families
- Al Gore, former vice president of the USA
  –Gore has become an influencer for a mixed-gender audience on environmental issues, with a particular focus on climate change
- Kim Kardashian, online influencer turned TV star
  –Initially, "famous for famous' sake" has evolved into a fashion icon for young adult women, especially between the ages of 19 and 32[4]

2. **Category Influencers**: They are people who have a genuine interest, expertise or enthusiasm in a topic (e.g., beauty, automotive, music). Usually, someone who is a practitioner in a particular area (ex: a professional chef who produces cooking videos) or has a certain involvement or expertise that makes other people believe that they are a reliable source of information related to that topic (ex: a passionate food blogger with hundreds of posts featuring her successful culinary creations). That person may not be famous among a broad audience, but they have a sufficient amount of "influence" that they can change people within their niche attitudes, beliefs, and behaviors.

Category Influencer can be broken down one step further into "Established" and "Emerging":

Established = A household name within a particular community
Emerging = A substantial, growing audience but not as widely known within a particular community

Examples of Category Influencers:

Topic: "Lifestyle Design"

- *Established*: Tim Ferriss, maven of "lifestyle design," author, and podcaster
- *Emerging*: Dan Andrews and his partner Ian Schoen, who operate the Dynamite Circle online community and Tropical MBA podcast dedicated to lifestyle design and location-independent entrepreneurship

Topic: "Beauty"

- *Established*: Michelle Phan, early YouTube make-up influencer who leveraged her online platform to launch ipsy, a subscription-based cosmetics company
- *Emerging*: Jacey Duprie, fashion and beauty blogger at "Damsel in Dior" with a growing audience of approximately 480,000 followers

Topic: "Entrepreneurship"

- *Established*: Gary Vaynerchuk, serial entrepreneur and acclaimed internet marketer
- *Emerging*: John Lee Dumas, host of the podcast interview series, *Entrepreneurs on Fire*, where he profiles successful entrepreneurs with an audience of approximately 100,000 followers

3. **Micro-Influencers**: These individuals do not necessarily have significant online reach (in fact, they may not even have an online presence), but they are very passionate about a specific topic, brand, or product category. It may be somebody who is championing organic food or somebody who absolutely loves Apple products, and they tell all their offline and online friends that they must buy these things.

Potential Examples of Micro-Influencers

- *Robert*: Avid car enthusiast without a social presence is the go-to person his friends, family and colleagues turn to when they need to buy a new car

- *Denise*: Casual "mommy blogger" with 2000 Instagram followers, but she receives multiple emails every week from blog readers asking her advice on what baby products to purchase
- *Alex*: Full-time graduate student with a passion for comic book collecting. His comic book collection-focused YouTube channel has less than 100 subscribers, but he stays up late at night responding to viewer comments with questions about how to best maintain the value of their own collections

Depending on a company's influencer-marketing strategy, it may engage with just one or all three levels of influencers, as evidenced by beauty company Glossier's rise to fame.

Roles of Influence: Different Personas for Different Use

"What's very motivating to us is this idea of every single woman being an influencer,[5]" explains Emily Weiss, founder and chief executive officer of Glossier, a beauty firm based in New York. Influencers are central to Glossier's marketing and product strategy; the company incorporates all three levels of Celebrity Influencers, Category Influencers, and Micro-Influencers throughout its product development life cycle.

The firm grew organically as an offshoot of Weiss' widely read beauty blog *Into the Gloss*, a side project she began while working as a styling assistant for *Vogue* in 2010. Her blog quickly developed a cult following due to her personalized approach to sharing beauty tips and the first-hand beauty advice of other ambitious women. She made a conscious effort to present beauty advice in an authentic, accessible manner, which ran in stark contrast to the traditional, authoritative editorial approach adopted by beauty magazines (like her then-employer) that presented an unattainable standard of beauty for everyday women.

In an ongoing interview series titled "The Top Shelf," Weiss (and now her editorial team) interview accomplished female professionals such as designers, artists, and fashion editors and produce intimate accounts of their beauty routines, career struggles, and feature pictures of their messy top-shelf bathroom vanities that display all the beauty products they use. The interviews offer a genuine behind-the-scenes glimpse into the everyday lives of ambitious women, an intimate experience otherwise reserved exclusively for their partner or close friends.

For four years, Weiss cultivated deep relationships with members of her online community to find out their likes, dislikes, and frustrations with their beauty routines. She discovered patterns and common needs that traditional, established beauty companies missed. Similar to how Tim Ferriss identified a problem with the traditional work–life dynamic and developed a community around his new way of thinking, Weiss used *Into the Gloss* to tap into the unmet beauty needs of millennial women.

By 2014, Weiss applied her years of market research and audience engagement to launch her first product line, which became the beauty company Glossier, whose aim is to provide everyday women with accessible luxury beauty products. Its brand purpose embraces everything Weiss learned during her four years of blogging and is highly mission-oriented:

> We're laying the foundation for a beauty movement that celebrates real girls, in real life. Glossier is a new way of thinking about (and shopping for) beauty products. Because "beauty" should be fun, easy, imperfect, and personal. Above all, we believe that you give life to products-products don't breathe life into you.[6]

As a privately held company, Glossier does not disclose official financial figures, but according to sources close to the firm, by the end of just the first year it recorded revenues in "multiple millions.[7]" In several interviews, Weiss said sales grew by 600% each subsequent year leading up to a cash infusion of $24 million in Series B funding by November 2016.[8] In 2018, the firm aims to go global, operating in its first set of international markets Canada, the UK, and France. What is behind the firm's rapid growth? Its dedicated community of Celebrity Influencers, Category Influencers, and Micro-Influencers.

Here are just a few examples of how Glossier collaborates with influencers:

Celebrity Influencers: Glossier interviews Celebrity Influencers for its "The Top Shelf" interview series, helping the firm gain third-party credibility and tap into the audiences of prominent influential women. Previous Celebrity Influencer interviews include Arianna Huffington, Kim Kardashian, and Jenna Lyons, former president and creative director of J. Crew.

Category Influencers: The Glossier public relations team regularly engages fashion and technology journalists and beauty bloggers to share the company story. Examples include:

Emily Weiss on What a Glossier Girl Smells Like and Building a Cool Girl Empire (Vogue)
How This Former Assistant at "Vogue" Is Building the 'Nike of Beauty' (Inc.)
How Glossier Hacked Social Media to Build A Cult-Like Following (Entrepreneur)

Category/Micro-Influencers: Glossier uses posts on *Into the Gloss* (now the official corporate blog), as well as the firm's Instagram account, to get real-time feedback about product development. When they asked the question, "Is this shade of lipstick too blue?" they received responses within moments, coming from Glossier Girl Micro-Influencers and established Category Influencers like celebrity make-up artist Mario Dedivanovic. Glossier also taps into this group to tease upcoming product launches to prime its referral sales program participants.

Micro-Influencers: Loyal customers discovered through social media can even get featured in new advertisements, blog posts, and Instagram posts furthering Weiss' vision that every woman can be an influencer. In her post announcing the Series B funding, she wrote, "Glossier is cult; it's not niche, and that's because we believe in the democratization of beauty. Glossier was created not to be for a privileged "some" but for an activated 'all'—and we are still early in our journey to fulfill that promise.[9]"

Glossier's story illustrates how in addition to "levels of influence" (Celebrity/Category/Micro) there are also different *"roles* of influence," which vary depending on the influencer's relationship with the brand. It's challenging to list all "roles" because some will vary from industry to industry, but the following is a selection of different roles of influence that companies may engage:

- Traditional journalists (e.g., fashion journalist)
- Industry experts (e.g., tech sector analyst)
- Academic (e.g., renowned professor on Topic X)
- Political (e.g., lobbyist)

Depending on what audience a brand is seeking to influence, the individuals they engage as influencers can be classified based on both their level of influence and role of influence. For Glossier, its team regularly engaged with traditional fashion journalists (role) for major media outlets. These same journalists have developed loyal followings of readers who seek out their knowledge as subject-matter experts on the latest fashion trends (Category Influencer—Level).

However, these are not the only factors that matter when it comes to understanding modern-day influencers, which is where the "influencer ABCC's" come in next.

Influencer ABCCs: Key Factors of Modern-Day Influencers

Regardless of the size of the audience they influence, or the role they play for companies that engage them, influencers must balance several key factors to be successful. It's time to learn your "influencer ABCC's."

A = Authenticity

An authentic, trusted relationship with a community is at the heart of what makes an influencer successful. When collaborating with brands, influencers' top concern is how to maintain this authentic connection, without being viewed as a "sell out." Before launching Glossier, Emily Weiss was an emerging Category Influencer for beauty. She could have written content to promote other cosmetics brands she didn't believe in just to make money, but she didn't. This would have been especially damaging to her credibility, if the products she chose to promote were outside of beauty, like iPhones or baby strollers (which happens all too often).

There are also legal considerations, especially for US-based influencers who need to be transparent about brand collaborations not just to help maintain authenticity, but also to protect against potential legal implications breaching Federal Trade Commission (FTC) regulations (see Chapter 10: *Know the Risks: The Dark Side of Influencer Collaboration*).

Carline Jorgensen, chief marketing officer of the integrated marketing agency Fanology, says, "'Authenticity' is a word that gets thrown around a lot, but it's truly at the heart of all of this. What we focus on is less on followers, but more on engagement, sentiment and business results."

Michael Krigsman, an industry analyst and host of CXOTALK, has a different perspective, "The best influencers don't lose their neutrality and independence. Where I'm being paid, I always disclose it. Even if I am being paid, I will not sell or pitch a company's product. I won't be their sales person—that's not my role. There is a compromise that needs to be made—based on having clarity around the mutual roles—what is the role of the company and what is the role of the influencer. As an influencer, you have to be extremely clear about the nature of the relationship you have with a given company."

B = Brand Fit

A consistent personal brand is critical for an influencer to gain more and more influence, but the influencer needs to balance their personal brand that sometimes comes at odds with the company brands they want to collaborate with. For Weiss, she could have easily partnered with an established beauty company like L'Oreal or Estee Lauder to help sell their existing products, or could have developed her own Glossier product line but sold it through a traditional distribution partner like beauty retailer Sephora. By choosing to go direct to consumers under her Glossier brand, Weiss ultimately maintained more control of her brand and strengthened her connection to her community.

Taryn Southern, a YouTube Category Influencer, explains, "The fact is, the influencer has their own brand, so when a brand collaborates with an influencer you basically have two brands working together with potentially different values and ways of engaging with their audience—which can sometimes make things messy." She argues that most influencers want to maintain the integrity of their personal brands as much as the brands that seek to partner with them. "As an influencer, you need to be able to see what is both in the best interest of your brand and the employing company's brand to best inform and design the collaboration. You have to be able to walk away if it doesn't feel right. Some influencers, particularly those who are still building, don't know how long they'll be able to sustain their success or audience, so there's a little bit of a mentality of needing to take on as much as possible right now. They feel like they can't say no. When they can't deliver, or compromise their brand to make money, no one wins."

From an agency perspective, Mark Fidelman, chief marketing officer of Fanatics Media, believes, "If we don't see a fit between a brand and an influencer, we don't facilitate the collaboration, because it will be obvious that the influencer is only participating for the money. The influencer's fans will know, and the influencer will be called out—that's a death sentence for an influencer. Their fans are not very forgiving."

C = Community (Reach, Resonance, Relevance)

A targeted, engaged, and growing community is the ultimate measure of success for an influencer. As a blogger, Weiss cracked the code to figure out how to acquire and develop the right type of community members that shared her passion for unfiltered, modern beauty. Influencer-marketing practitioners often refer to the three factors used to measure an influencer's community as their Reach, Relevance, and Resonance:

- *Reach*: The total size of an influencer's audience across all social platforms measured by followers, subscribers, traffic, etc. Think of reach as what defines an individual as a Celebrity Influencer, Category Influencer, or Micro-Influencer, as related to the earlier section of this chapter.
- *Resonance*: Engagement between the influencer's audience and the content they produce measured by shares, likes, views, comments, retweets, etc. Resonance is important, because influencers need to be able to demonstrate that their community is more than just a number that it's engaged and interested in their content. Remember the saying, "if a tweet is sent to 100,000 followers and nobody retweets it, did the tweet have any resonance at all"?
- *Relevance*: Content-topic match ensuring that the content produced by the influencer (see next "C" below) is aligned with a consistent set of topics that is of interest to the influencer's community. From a brand's perspective, relevance also relates to how closely an influencer's community matches up to the brand's target audience, as well as how closely the influencer's content aligns with key topics that the brand wants to be associated with.[10]

"If somebody is truly influential for something, then they've earned their audience and the engagement shows," explains Matt Britton, chief executive officer of CrowdTap. "It's not just the amount of comments it's are people commenting about the specific topic. If somebody is influential for making sushi, then are the comments responding about how to make sushi. That shows that the influencer has gained influence in that specific topic."

Delphine Reynaud, vice president of influence at influencer-marketing software firm Traackr, makes the case, "Sometimes I speak to an influencer who has 250,000 followers, but when I use software to analyze their audience demographics, I'll find that his followers are trivial people in developing countries who are interested in other topics. In general, it's really about quality, not quantity."

C = Content

Content is how an influencer adds value to and builds a relationship with their community. Which leads to another ongoing challenge influencers face: how to come up with new content—not just any content, but content community members can't find anywhere else delivered in creative ways that continue to engage their community within a consistent set of topic/focus areas.

It's common to hear people refer to influencers based on the type of platforms the influencer uses. For example, "YouTubers" or "bloggers," but the reality is all forms of social media are central to the emergence of the modern-day influencer (see Chapter 2). Social media provides influencers channels to share their content in a manner they're most comfortable with (not everyone is camera-ready, conversely not everyone is an eloquent long-form writer). In addition to sharing content, social media helps influencers drive two-way engagement with a growing audience that amplifies their influence over time.

It's fairly common for influencers to appear across multiple platforms. For example, bloggers generally use platforms like Facebook and Twitter to share their posts. However, most influencers have a go-to platform which leads to classifications like YouTubers, Bloggers, and Instagrammers.

Here is a high-level selection of some of the content/platform choices influencers may work:

- Recorded Video—YouTube
- Live-Streaming Video—Facebook Live
- Photographs—Instagram
- Audio—Podcasts
- Long-Form Text—Blogs
- Short-Form Text—Twitter
- Curated Content—Pinterest
- Professional Advice—LinkedIn

Weiss' consistent use of her blog, *Into the Gloss*, and Instagram, combined with an evolving set of creative content about modern, real-life beauty, enabled her to engage and grow a community that ultimately launched a multi-million-dollar beauty company in the form of Glossier.

"Any type of content needs to be designed to trigger behavioral change—to move people along on a journey from noticing to learning more, to buying more to advocating more. With each one of these journey points you end up with greater loyalty," explains Tom Doctoroff, senior partner at Prophet.

Meanwhile, brands oftentimes do not appreciate the effort that goes into producing high-quality content. Heidi Nazarudin a fashion Category Influencer and founder of TheAmbitionista.com shares her experience: "At a recent shoot that I did for a cosmetics company, I had to pay the photographer, I had to pay a lighting person, had to hire a driver, had to go to a studio…the cost of the shoot alone—production cost—is about $5000. Brands will come to me and say 'we love the campaign you did for brand X' and when I tell them the necessary budget they step back. I know a lot of bloggers deal with this—Will you help us for a box of our products? No."

Without the A-B-C-C combination, it's impossible for someone to truly have influence. If someone has a few hundred thousand followers but they pitch different brands every single day, then they'll lose their audience (lack of authenticity/brand-fit). Alternatively, if someone produces great content, but has a community of zero readers, then…you get the point. These four factors combined are what truly makes an individual influential in the modern-day technology-enabled influence landscape.

Finally!!! A Common Definition of Modern-Day Influencers

In Chapters 2 and 3, you met several influencers, ranging from YouTuber Lilly Singh and car enthusiast Robert to author/podcaster Tim Ferriss and fashion blogger Emily Weiss. Marketers unfamiliar with the ins and outs of influencer marketing may find it confusing when blanket terms like "influencer" and "Instagrammer" are used to talk about anyone with an online following. Who are these influencers?

First, influencers can be classified based on the size of the communities they have influence over. Recognizing that follower numbers can be exaggerated, if companies apply lessons to measure influencer quality in Chapter 7: *Discover Influencers: Finding the Perfect Match*, fake followers will not be as much of an issue. Grouping influencers as Celebrity Influencers, Category

Influencers (both emerging and established), and Micro-Influencers brings order to an otherwise confusing landscape where everyone throws around the all-encompassing term "influencer" to describe everyone from Robert the car enthusiast to Kim Kardashian.

However, having celebrity status or a large online following does not make them influencers. To be considered an influencer, the individual needs to be (i) *known for something* among a target community and (ii) able to *influence the actions* of target-community members. Being famous for famous' sake does not an influencer make.

Beyond their community size, influencers also play a variety of roles when it comes to the companies they collaborate with. Most often when it comes to influencer marketing, people envision a pay-to-play influencer endorsement post (hotel pays an influencer to post pictures from their resort on Instagram). But the ways influencers and brands interact are much more substantive (don't miss Chapter 9: *Working with Influencers: Potential Paths to Take*). Celebrity endorsers, product development informers, brand advocates, customers, employees—depending on the use case, anyone from a Celebrity Influencer to a Micro-Influencer can all play significant roles in a company's influencer-marketing strategy—just look at what Glossier accomplished in under four years with its influencer-led approach.

When engaging influencers, companies also need to keep in mind their influencer ABCCs. Structure engagements to help influencers maintain their authenticity, not conflict their personal brand, add value to their community, and create new, exciting content are all foundational to successful collaboration.

For this book to focus only on North America-based influencers would be a missed opportunity. As fast as the space is developing in the USA, the borderless nature of social media and online platforms has led to the emergence of a new set of Celebrity Influencers, Category Influencers, and Micro-Influencers in overseas markets around the world. These Local Influencers present a unique opportunity for companies from outside the influencer's country to tap into their deep understanding for their local community and local business culture.

How is influencer marketing developing around the world and what opportunities exist for marketers seeking to leverage influencer marketing to reach new international audiences?

Turn to Chapter 4: *A Global Phenomenon: The Rise of Influencers Around the World*.

Marketer's Cheat Sheet

- **Levels of Influence**: While people try to simply call everyone with an online community an "influencer," it's more informative to first break down influencers by their audience size to classify them as either Celebrity Influencer, Category Influencer (emerging or established), or Micro-Influencer and normalize audience size to account for fake followers.
- **Follow Count ≠ Influence**: Fame and follower count alone does not classify an individual as an influencer. The individual needs to be (i) *known for something* among a target community and (ii) be able to *influence the actions* of target-community members. Being famous for famous' sake does not an influencer make.
- **Roles of Influence**: Depending on how an individual engages with a company, they can assume different roles of influence. Glossier illustrates how influencers with various-sized audiences can be incorporated throughout the product-development life cycle to deliver meaningful business impact.
- **Influencer ABCCs**: When seeking to collaborate with influencers, companies need to take into account the influencer's perspective, especially related to authenticity, brand fit, community (Reach, Resonance, Relevance), and content.
- **Authenticity**: An authentic, trusted relationship with a community is at the heart of what makes an influencer successful. When collaborating with brands, influencers' top concern is how to maintain this authentic connection, without being viewed as a "sell out."
- **Brand Fit**: A consistent personal brand is critical for an influencer to gain more and more influence, but the influencer needs to balance their personal brand that sometimes comes at odds with the company brands they want to collaborate with.
- **Community**: A targeted, engaged, and growing community is the ultimate measure of success for an influencer.
- **Content**: Content is how an influencer adds value to and builds a relationship with their community.

Notes

1. Ferriss, Timothy. *The 4-Hour Workweek*. New York: Harmony Books, 2009, 12–16. Print.
2. BusinessDictionary. "What Are Influencers? Definition and Meaning." *BusinessDictionary.com*, www.businessdictionary.com/definition/influencers. html.
3. Most companies make the mistake of choosing to work with influencers based purely on the size of their audience (fans/followers/etc.). This is a HUGE mistake. There are many morally-questionable practices some influencers employ to artificially increase their audience statistics (see Chapter 10— *Know the Risks: The Dark Side of Influencer Collaboration*). The three

"levels of influence" introduced in this chapter assume a vetted audience size that has not been inflated by dishonest practices.

4. *HYPR Influencer Database*, 1 Nov. 2017, "Kim Kardashian."

5. Avins, Jenni. "Glossier Is Building a Multimillion-Dollar Millennial Makeup Empire with Slack, Instagram, and Selfies." *Quartz*, 1 Dec. 2016, qz.com/847460/glossier-girls-emily-weiss-on-how-glossiers-customers-became-its-most-powerful-sales-force/.

6. Glossier. *Glossier About Page*, www.glossierGlossier.com/about.

7. Avins, Jenni. "Glossier Is Building a Multimillion-Dollar Millennial Makeup Empire with Slack, Instagram, and Selfies." *Quartz*, 1 Dec. 2016, qz.com/847460/glossier-girls-emily-weiss-on-how-glossiers-customers-became-its-most-powerful-sales-force/.

8. Weiss, Emily. "How We Raised Our Latest Round of Funding." *Into the Gloss*, Glossier, Nov. 2016, intothegloss.com/2016/11/glossier-series-b-funding-announcement/.

9. Weiss, Emily. "How We Raised Our Latest Round of Funding." *Into the Gloss*, Glossier, Nov. 2016, intothegloss.com/2016/11/glossier-series-b-funding-announcement/.

10. Solis, Brian. *Influence 2.0: The Future of Influencer Marketing*, 2017.

4

A Global Phenomenon: The Rise of Influencers Around the World

Elijah Whaley will never forget the early difficulties of collaborating with his girlfriend, a Chinese beauty influencer named Maggie Fu, on one of their first product sales promotion campaigns: "I cringe just thinking back to it – in a matter of days we went from a serious high, selling $25,000 worth of products, to an all-time low, falling into $25,000 of debt. I didn't think we were going to be able to bounce back," he says.

Whaley, an American marketing professional from Nebraska, accepted a friend's invitation to travel to China to work in the fast-paced tech start-up scene there. In 2015, he met his girlfriend Maggie, a Hong Kong Chinese, during a swing dance class in Beijing. He was smitten at first sight; it was more than just her physical beauty (she started modeling in Hong Kong as a teenager)—she possessed an intangible charisma that drew in not only Whaley, but anyone who came in contact with her.

Shortly after the couple began dating, Whaley learned of Maggie's aspirations to launch her own cosmetics brand. She graduated from Hong Kong University with a degree in fine arts and worked as a makeup artist, building up a niche online following through writing guest posts on Chinese beauty and lifestyle blogs in her evenings. When Whaley saw that Maggie's guest articles had earned her over 20,000 followers on the Chinese Twitter equivalent Weibo (popular global social media sites like Twitter, Instagram, and Facebook are not available in China), he suggested that Maggie begin producing online makeup tutorials to continue to grow her audience, and ultimately use her engaged following to launch her own cosmetics brand.

© The Author(s) 2018
J. Backaler, *Digital Influence*,
https://doi.org/10.1007/978-3-319-78396-3_4

Out of that conversation, "Melilim Fu," Maggie's online persona, was born.

The couple combined their professional strengths to build the Melilim Fu brand. Whaley managed the business and technical side of the influencer relationship, focusing on activities ranging from filming and editing videos, developing pitch decks, and marketing materials and contacting potential brand partners. Meanwhile, Maggie spent her time experimenting with different content formats and engaging fans in small messaging groups to get their feedback on their likes and dislikes and refine her content production approach. Their relentless hustle (all of this was happening while they both worked demanding full-time jobs), along with a few opportunistic breaks being featured in Chinese online media, led to Melilim Fu having collaborations with major international brands H&M, Nike, and Estée Lauder within the first year.

Right around their one-year anniversary in spring 2016, a friend of a friend approached the couple with an offer to collaborate on a luxury handbag promotion. They quickly struck a deal, and Melilim Fu spent seven days promoting the bags across her social media channels. Through Melilim Fu's trusted online relationships with her audience, in less than seven days she sold $25,000 worth of purses.

Even more so than in the USA, Chinese digital consumers are willing to quickly make purchases based solely on the recommendations of influencers they follow. Whaley couldn't believe how effective Maggie's promotion was. "It was like all of a sudden, we had this incredibly valuable sales channel, and we were just getting started," he says, thinking back to the moment the final sales numbers came in.

But then, when the duo received the first shipment of bags, something went terribly wrong—the "luxury bags" were fakes, and there was nothing they could do about it. It was their fault for not doing their homework in advance, like requesting product samples to inspect quality. Based on their agreement with the Chinese partner, they still had to buy the $25,000 worth of bags that were purchased on consignment. Maggie was devastated, feeling like she broke her audience's trust. She immediately refunded the $25,000 in orders to her followers who purchased the bag. In an instant, an initial win became their worst influencer collaboration experience.

Thankfully, Melilim Fu reestablished trust with her audience and has recovered from this experience. At the time of writing this book, she has more than 500,000 Weibo followers, has hundreds of successful brand collaborations under her belt, and is a full-time beauty influencer, laying the foundation for her future cosmetics brand. Much of her success has been due to factors that are consistent to influencers in other markets, but some, as you are about to learn, are related to unique local market nuances.

A Global Phenomenon: Influencers Emerge Around the World

Chapter 2 introduces how rising technologies and disappearing gatekeepers enabled the ascent of the modern-day influencer—content used to be very expensive for an individual to produce and promote, creating content required specialized knowledge and expertise, and promoting content to broad audiences needed approval from industry gatekeepers. This chapter explores another key barrier that is becoming less and less relevant, geographic barriers.

The borderless nature of the Internet, as well as widespread international adoption of global social media (including Facebook, LinkedIn, Instagram, Twitter, and YouTube), means almost everyone has access to the same information in real time. The only limitations are discovery (being able to find out what's available with so much content online), language (being able to consume the content), and potential local market nuances (like government restrictions on certain websites or content). This means with the right audience engagement strategy, influencers from one country can build a following in far-flung locations around the world where they have never even traveled. On that same token, marketers can now collaborate with Local Influencers in foreign markets to reach new overseas audiences in an authentic manner using a localized, culturally sensitive approach (Fig. 4.1).

The global influencer marketing landscape is evolving quickly, but differently, around the world. Maddie Raedts, founding partner and chief creative officer at the Netherlands-based influencer marketing agency IMA points out, "The United Kingdom and the United States are the most saturated markets, since they were the first movers in this space. Outside the UK, in developed European markets there is a tremendous amount of diversity, even between countries that are close to one another. But, there is a sense of fluidity as well, with influencers having pan-European followings through being able to travel to these varied landscapes".

Raedts also sees a lot of potential for influencer marketing in emerging markets across Eastern Europe and Asia. "The engagement rates of influencers

"This new era of connectivity has companies considering global expansion much earlier than they once did. Local partners and influencers hold the key to opening new markets faster than ever before."

-Joel Whitaker (@joelwhitaker)
SVP, Research
FRONTIER STRATEGY GROUP

DigitalInfluenceBook.com   SHARE THIS

Fig. 4.1 Joel Whitaker

in Russia, for example, are seven times higher than the average engagement rate in Europe." However, when it comes to Asia while she is optimistic about growth prospects in the region, she also recognizes that it is more difficult for American and European companies to consider due to vastly different languages and several popular social media platforms distinct to specific countries in that part of the world.

Meanwhile, in the Middle East, there are no big agency players that have emerged to manage influencers and brands are still learning how to work with influencers. "It's still something new here, which means that most local brands are not yet aware of its importance (some brands don't even know that influencers get paid to create content) so a lot of local brands still prefer to advertise their products on TV rather than through influencers – but this is changing fast," explains Soukaina Aboudou, a Middle East-based beauty influencer.

In Latin America, Celebrity Influencers remain the top choice for marketers seeking to reach a broad regional audience, because outside of Mexico and Brazil it is not practical to take a market-by-market approach, as most Latin American countries are generally too small to justify the dedicated influencer marketing spend. However, in order to work with brands, most Latin American Celebrity Influencers are still tied to gatekeepers within the traditional studio system at major media giants.

Xavier Mantilla, head of Digital Market Media and former Latin America regional director for Starcom MediaVest Group, explains the bind Celebrity Influencers are in: "The networks themselves, Grupo Televisa and Grupo Globo, tell their talent 'if you're under contract to me as television talent, then your social media belongs to me.' If the Celebrity Influencer says 'no' then the media company can prevent them from getting future television show opportunities and essentially kill their career."

In Africa, across the continent "the market remains to a degree immature, largely due to technical constraints," argues Nicky Schermer, managing director of Ogilvy Public Relations in Cape Town.[1] For the time being, this limits the availability of local Category Influencers with sizeable audiences. However, the widespread popularity of Nigeria's burgeoning movie production industry known as *Nollywood* (second biggest domestic industry only to farming in the continent's largest economy) has given rise to a host of Celebrity Influencers with impact across the continent. Or, as one local put it, "Nigerians have succeeded through Nollywood to export who they are, their culture, their lifestyle, everything" (Fig. 4.2).[2]

Then, there is massive potential in markets with large populations that are highly socially active online like Brazil, India, Russia and China. "Much of

"The rise of influencers extends well beyond geographic borders, and is relevant across industries – both B2C and B2B alike. This is a global phenomenon."

-Josh Steimle (@joshsteimle)
Founder
MWI & INFLUENCER INC

DigitalInfluenceBook.com   SHARE THIS

Fig. 4.2 Josh Steimle

that is based upon the size of their online populations and how active they are on social networks," clarifies Christian Damsen, a senior vice president at the influencer marketing software firm Traackr. "Taking it a step further, the amount of e-commerce and purchases that are being made online through influencer activations with their audience in several of those markets is incredible."

In China, for example, the social media platform WeChat, in addition to being a content distribution channel for influencers to share articles with their audiences, also has digital wallet and micro-site online store functionality, which makes it possible for influencers to seamlessly introduce products to their audience and drive followers to make purchases within the same mobile app. This "ease of e-commerce" is a major factor that led Melilim Fu to sell $25,000 worth of fake handbags in one of her first online promotions. It also led Tao Liang, also known as "Mr. Bags," an established Chinese Category Influencer, to sell 1.2 million RMB ($173,652) worth of genuine Givenchy designer handbags in just 12 minutes.[3]

For marketers, collaborating with Local Influencers in overseas markets is not for the faint of heart. As Chapter 9 will reveal, working with influencers even in your home country can be a complicated endeavor, but the added dimensions of a different country, culture, time zone and often language can make collaborating with Local Influencers all the more complex. Despite these challenges, companies that crack the code to successfully engage Local Influencers can add an invaluable channel to their international expansion strategy ranging from overseas sales to product development, brand awareness, and a wide variety of global corporate initiatives.

That being said, as with anything related to doing business in a foreign market, cultural understanding and the ability to adapt to local market nuances is at the foundation of a successful strategy. A simple "translate and transplant" (i.e., slap a foreign label on what you do in your home country and hope to achieve similar results overseas) is a recipe for disaster. Andy Molinsky, a professor of international management at Brandeis University, contends, "Whether we're talking about influencer marketing or any type of marketing, when it comes to doing business overseas, cross-cultural understanding is fundamental. Marketers need to take the time to understand the local cultural context that their target influencers operate in, and adjust their approach accordingly." Both understanding the local cultural context and how Local Influencers can help foreign brands operate more like locals is critical.

Cultural Influence: Bridging the Culture Divide to Reach Local Audiences

When executed well overseas, influencer marketing can help a company appear less "foreign" and allow a brand to connect with local audiences in a culturally-relevant manner. However, depending on your industry, influencer marketing may not be the best fit for every international market you do business in; there are certain cultures that are naturally more receptive to an influencer-led marketing approach.

For example, consumers in most Asian cultures tend to look for assurance that a product is going to work before making a purchase decision. They do everything possible to avoid "losing face" (publicly make a mistake and face societal embarrassment). "In general, Asian consumers are more 'reassurance driven' relatively speaking and they trust less," explains Tom Doctoroff, senior partner at Prophet. "In any hierarchical society, you're going to have a need to not lose face, and ultimately not having institutions that are as reliable as they are in the West (ex: to ensure food safety, product quality standards, etc.) only strengthens the need for the third-party validation that trusted influencers may offer."

This reassurance-driven approach to decision making holds true for businesses who sell to other businesses (B2B) in certain Asian markets as well. For example, American and European technology companies have a long-established history of selling indirectly through local third-party firms to sell products to Japanese companies. Executive decision makers at Japanese companies put a lot of faith in the local expertise of the Japanese Category Influencers who represent newer, lesser-known technology products developed by foreign companies. Rather than setting up their own local Japanese operation, the American technology company DocuSign adopted this approach. In 2016, DocuSign established a software reseller agreement with Mitsui Knowledge Industry Co., Ltd. (MKI), due to MKI's long-term presence in the market, and experienced team of local representatives (Fig. 4.3).[4]

Latin America is another region where influencer marketing may prove more effective than others, especially for consumer-focused firms. Xavier Mantilla remarks, "The big thing about Latin America is you have a large group of people who are aspirational consumers. I think that won't go away, because class structure in Latin America is very clear – if you're born into the right family you have a great life. If you're born into a family that doesn't have much, there's a lot more struggle. Most countries do have a middle class that has grown substantially so you have a lot of consumers

"Local influencers help bridge the cultural divide between foreign brands and new overseas audiences. They're an essential component to international expansion."

-Heidi Yu (@heidihanyu)
Founder & CEO
BOOSTINSIDER

DigitalInfluenceBook.com   SHARE THIS

Fig. 4.3 Heidi Yu

that are very aspirational." For influencer marketing in Latin America, this means there is still a lot of white space for brands who target aspirational consumers.

Then, there are some cultures where influencer marketing—or any form of marketing for that matter—for certain industries cannot effectively reach local audiences. For example, promoting any alcohol or tobacco in the United Arab Emirates (UAE) is forbidden. Companies also need to be aware of if Local Influencers comply with local laws. According to UAE laws, influencers require a trade license to promote products and accept payment for their social influence. While this law may not be strictly enforced today, as the use of influencers grows and becomes more prominent, it will become increasingly important for companies to be aware if an influencer has the appropriate corporate structure in place.

Marketers need to do their local market research to understand if a foreign culture will be receptive to influencer marketing in their industry before simply copying and pasting a strategy that has proven successful in their home country. As with any other set of influencers, Local Influencers can be categorized as Celebrity Influencers, Category Influencers, and Micro-Influencers. However, there are two additional ways to categorize foreign influencers based on whether their audience is based in a single foreign market or multiple foreign markets.

Global Influence: Where in the World Do Influencers Have Influence?

Xenia Tchoumi embodies the phrase "Global Citizen." Born in Russia, raised in Switzerland, and currently residing in London, Xenia operates across cultures and languages (she speaks six of them: Italian, Russian, German, French, English, and Spanish). Although she is a fashion Category Influencer and former model, don't mistake her for just a pretty face: Xenia holds a degree in economics, has previous professional experience at JP Morgan Chase, and does high-profile speaking engagements at the United Nations, as well as multiple TED talks. She shares updates on the latest international travel and fashion trends through her blog *Chic Overdose* and via social media content primarily published on Instagram and Facebook.[5]

While Xenia is technically based in London, she might not be a good choice for companies seeking to collaborate with her on UK-specific campaigns. That's because most of her audience is not in the UK. Given her international background, and regular travel schedule, Xenia has built a following in many markets outside her home country. The top five geographies where her audience is based are the USA, Switzerland, UK, France, and Italy.[6]

This breakdown of Xenia's audience illustrates why marketers, in order to understand if an individual is a good influencer to reach a local audience, must consider much more than where an influencer is based—they need to understand where the influencer's *audience* is based. The idea is it's not necessarily where the influencer him/herself is based, but where their audience is based which makes them attractive for a particular brand. That's why in addition to categorizing influencers as Celebrity Influencers, Category Influencers, and Micro-Influencers, when doing business internationally, marketers should determine whether an influencer is Local (the majority of their audience resides in their home country) or Global (the majority of their audience is distributed across multiple geographies).

Local Influencers

Tuka Roberta Rossatti, media manager at Havas Group based in Rio de Janeiro, feels that influencers in Brazil primarily produce content that serves two purposes: "Either they translate what's happening in the world or what's trending globally for Brazilians, or they translate ourselves – our own regional local cultures for Brazilians across the country to learn from."

Whether in Brazil or other countries, Local Influencers open the door to their home market. Their audience is local, and they produce content that is relevant for a local audience. If someone is not from that local market, they likely will not understand the content, either due to cultural nuances or plainly not being able to understand the local language.

Some examples of Local Influencers include:

- Brazil: "whinderssonnunes," Brazilian YouTube comedian influencer with approximately 23 million subscribers[7]
- China: "DaddyLab" (老爸评测), a Chinese parenting blogger who tests products to make sure they meet required safety requirements with more than 120,000 WeChat followers[8]
- Russia-Ukraine: Ivan Rudskoi, "EeOneGuy," a Russian-Ukrainian gaming influencer with approximately 2.4 million followers on the Russian social media site VK.com, 12 million followers on his YouTube channel[9]

"The reality is we've been speaking about key opinion leaders for as long as 20 years," says Bruna Scognamiglio, vice president global influencer marketing for Gucci Beauty. What has changed, in her opinion, is the way these influencers act—they used to be involved mainly in live events or in traditional media like fashion magazines; now it's more digital. Additionally, in the past there were few influencers who were considered truly "global," and they tended to be people who were considered very influential based on their specific business/industry.

"Now you still have these global, high-level people," Scognamiglio says, "but you also have the voice from the local-level that can be even more important than the global players – people connect more to their Local Influencers – the millennial generation is extremely interested in authentic messages. That's why the voice of people who are authoritative and have their own point of view on a special topic is where millennials will go to listen who are interested in that topic – they won't necessarily only turn to a mainstream global influencer" (Fig. 4.4).

This insight is extremely important for the makeup business, which has always been driven by the experiences and the recommendations of peers. "A typical consumer goes online in their own language, goes on a platform to check who the makeup bloggers are, they watch tutorials, try the products and listen to their advice," Scognamiglio explains, "These consumers

"Influencers with followings concentrated in specific geographies help brands reach their target local audiences in an authentic manner. That's exactly how I work with travel brands to reach Chinese consumers."

Lauren Hallanan, China-Global Travel Influencer

DigitalInfluenceBook.com   SHARE THIS

Fig. 4.4   Lauren Hallanan

don't usually look at big Global Influencers because they think their advice isn't necessarily compatible." She offers the examples: depending on the skin tones, traditions or habits the way of applying makeup may be very different and hence the influencers women refer to may be more local than global. For instance, in the Middle East "the kind of makeup she uses is quite different and they focus more on the eyes while for instance in Asia there is big attention to get the perfect color for the base skin."

But there is a role for Global Influencers as well, as companies look to tap into influencers with audiences in multiple markets.

Global Influencers

There may be cases when brands will intentionally approach influencers outside their home country who are able to influence audiences in locations around the world. Global Influencers appeal to multiple audiences in multiple geographies, often due to commonalities like shared languages, similar cultures, natural industry location concentrations. It boils down to this: there's something about a Global Influencer that makes them less foreign, and thus appealing to audiences in certain countries. Oftentimes, it's not clear exactly why this is the case.

In speaking with Tim Williams, chief executive officer of Onalytica, an influencer marketing software firm, he explains, "You need to figure out who is influencing globally versus who is influencing locally." Based on his firm's research they found, "If you look at English, Spanish, French you can start seeing where the influence is by language – sometimes its localized in a market, sometimes it could be people in Argentina influencing Spain. It's really interesting to analyze because it doesn't always fit with how organizations are generally structured by geography."

Meanwhile, Beca Alexander, founder and president of the influencer marketing agency Socialyte, points out, "Most of our influencers are what we call 'global citizens,' which means they have a global following." Alexander feels that there are a lot of American brands that are looking for an influencer who reaches a large, global following across different countries. "We run many campaigns where an American brand is opening a store in London and we need to find London-based influencers who would promote the store opening, or an Australian brand opening a store in New York City where we need to find an influencer who can reach both audiences."

Some examples of Global Influencers include:

- Influencer: "Fer0m0nas," a gaming influencer with 3.3 million YouTube subscribers

  - Influencer Location: Portugal
  - Audience Location: Brazil
  - Global Connection: The shared language of Portuguese enables Fer0m0nas to connect to the Brazilian-based gaming community

- Influencer: Trevor Noah, host of Comedy Central's "The Daily Show" political influencer

  - Influencer Location: USA
  - Audience Location: South Africa
  - Global Connection: Originally from South Africa, the majority of Noah's audience is still based there

- Influencer: Ronald van Loon, top B2B influencer for Artificial Intelligence, Big Data, and Internet of Things

  - Influencer Location: the Netherlands
  - Audience Location: USA, Europe, India
  - Global Connection: The industries and topics covered by Ronald have strong followings outside his home country of the Netherlands

Availability of Local and Global Influencers

Marketers need to be aware that they may not have as many choices of specific types of influencers in certain countries. When interviewed for this book, Ronald van Loon discusses how American B2B companies usually turn to him for the European perspective: "The funny thing is, even if an American company holds an event in Europe, in the data and analytics domain, it's Ronald and the US people."

He feels that there is a limited number of European influencers in his niche. "I focus almost exclusively on the US – 90% of my contacts and deals come from American companies or the US subsidiary of European companies," he says. He attributes the lack of European influencers writing about new technologies to the fact that many Europeans come from small countries, and they often do not regularly share content in the English language; however, he notes that this is changing.

Huawei, the Chinese technology company, has similar experiences while building out their global influencer network. Huawei manages a community of 100 influencers from a diverse range of markets: Australia, Malaysia, India, Pakistan, Saudi Arabia, UAE, Nigeria, South Africa, Poland, Russia, Turkey, France, Ireland, Canada, USA, and Mexico. Walter Jennings, head of Huawei's influencer relations, has a similar experience to van Loon's above:

"What we found is in identifying influencers in many of these markets is we had to broaden the types of individuals we included. You're not going to have a pure play B2B influencer—you're probably going to have a more general 'tech influencer.'" In addition to a broader area of focus, these influencers likely have a much smaller following, but "they're really well-respected in their home country as well in their regions." Jennings offers the example of his influencer in Pakistan, Muhammad Saad Khan, who at the time of writing this book only has 3700 followers, but he is very well respected in both the local Pakistani technology community and the neighboring Middle Eastern markets.

Opportunities to Go Global While Staying Local

Yuping He leads China digital marketing for ZooEnglish.com,[10] an American online education company specializing in early childhood education. After years of successfully operating in the USA and Europe, ZooEnglish's management set out to expand into the Chinese market. Shortly after their decision, they hired Yuping, a Chinese-born American citizen, to build an initial digital strategy for their local online presence across the Chinese social media platforms WeChat and Weibo.

The major catch is, management made it clear to Yuping that they have no intention to open a local Chinese subsidiary, and they want the entire China operation managed centrally at their headquarters in California. With over a decade of Chinese digital experience, Yuping realized she had her work cut out for her—almost every company she knew that was successful in China had a local office there—but she came up with a creative solution to satisfy her management's need for control.

She realized that California is home to one of the largest overseas Chinese populations in the world. San Francisco and Los Angeles alone are home to more than one million Chinese immigrants (not to mention thousands of Chinese students and tourists). Yuping began to build online relationships with Chinese childhood education influencers based in California. She spent weeks combing through profiles, reading their content, assessing follower engagement levels and reviewing comments to select a handful of high-quality US-based Chinese influencers. After establishing a virtual relationship, Yuping began to drive across the state to meet with each of her top influencers in person. Eventually, she invited her top three influencers to ZooEnglish headquarters to meet with management and learn more about their company.

This respected management's need for control and gave Yuping the green light to partner closely with her vetted group of influencers. It was the start of an extremely fruitful influencer relations program—in the first 24 hours of collaboration, one of her influencers sold $100,000 worth of product licensees that retail for approximately $100 per year. The influencer relations program has become a core component of ZooEnglish's China market strategy.

Marketers should learn from the ZooEnglish example above and recognize that there may be opportunities to tap into the diaspora populations in your home market to target culturally similar audiences in overseas markets. The same approach Yuping used for the Chinese market could be deployed by marketers seeking to reach audiences in Latin America, India, the Middle East, and many other markets with large diaspora populations spread throughout the world.

Going Global: Influencers' Expanding Role in International Marketing

Chapter 3 finally made sense of modern-day influencer marketing and introduced the three influencer levels of Celebrity Influencers, Category Influencers, and Micro-Influencers. Chapter 4 expands upon these definitions to demonstrate how influencers and influencer marketing are coming of age in countries all around the world due to the borderless nature of social media platforms with international users.

First, influencer marketing is evolving at different rates across around the world: The USA and UK are often referred to as the most developed countries for influencer marketing, while other markets in Europe are quickly catching up. Local market nuances mean certain countries are seemingly more advanced in particular areas like e-commerce: Chinese WeChat influencers possess an uncanny ability to get their followers to make immediate purchase decisions based on their recommendations. Meanwhile, other markets in Latin America and Africa are still catching up because many of the barriers introduced in Chapter 2 (industry gate-keepers, under-developed technology) continue to persist.

Second, as with any sound international marketing strategy, cultural considerations are key. Marketers must research local trends and regulations to determine if influencer marketing is the right approach in a given market for their industry and specific product offering. No matter how much money a liquor brand is willing to throw at influencer marketing in the UAE; it's not the right approach for the market.

Marketers can choose to engage influencers who maintain an audience in a single market (local) or influencers with sizeable groups of followings in multiple markets (global). Regardless of whether a brand chooses to work with a Local or Global Influencer, it needs to conduct deep research to ensure the influencer's audience does in fact match the target audience that the brand seeks to reach.

Finally, when collaborating with Local and Global Influencers, marketers should be open to experimenting with different approaches. While a country may not have the exact type of influencer you work with in the USA or UK, you may be able to identify an influencer with a slightly broader category expertise who could still advance your goals in the local market. Alternatively, marketers should be willing to think more holistically about how to influence consumers in an overseas market—the ZooEnglish case study illustrates how it may be possible to "go global" while staying local.

Beyond geographical differences, what are some of the ways that influencer marketing is developing differently across industries? The next chapter introduces what you need to know about the business to consumer (B2C) influencer marketing landscape.

Marketer's Cheat Sheet

- **A Global Phenomenon**: Influencer marketing is expanding in countries all around the globe. The rise of new technologies and borderless nature of social media platforms have fueled an international movement.
- **Development Varies**: Even though there are "influencers" in most major markets, the level of maturity varies from country to country. Don't expect that a market in Europe or Asia will have the same talent, agencies, software, etc., that you may be accustomed to in a market like the USA or UK.
- **Cultural Context is Key**: Cultural nuances will play a significant role in any international marketing strategy that taps into Local Influencers. Certain markets like Asia may be more receptive to the third-party endorsements from trusted influencers, while influencer marketing efforts in other markets may be met with greater skepticism or local regulatory hurdles.
- **Local vs. Global Influencers**: Depending on a brand's objectives, it may choose to incorporate Local or Global Influencers into their international marketing strategy. Regardless of which they choose, marketers must take a deep dive into the influencers' audiences to ensure they match the brand's target audience.
- **Influencer Scarcity**: Because influencer marketing is developing at different rates around the world, not all countries will have the exact type of influencer you may be used to collaborating with. Keep an open mind and consider partnering with a local category influencer with a slightly broader focus.
- **Global Mobility**: With diaspora populations popping up around the world, local in-country influencers may not be the only option. Marketers can determine if there are influencers from Country X-based where their company is headquartered, to build an online and offline relationship that ultimately allows the brand to reach overseas customers in a more controlled manner.

Notes

1. Toesland, Finbarr. "Africa's New Media Influencers." *African Business*, 2 Feb. 2017, africanbusinessmagazine.com/sectors/technologytechnology/africas-new-media-influencers/.
2. Onishi, Norimitsu. "Nigeria's Booming Film Industry Redefines African Life." *The New York Times*, 18 Feb. 2016, www.nytimes.com/2016/02/19/world/africaAfrica/with-a-boom-before-the-cameras-nigeriaNigeria-redefines-african-life.html.
3. Chitrakorn, Kati. "How China's 'Mr. Bags' Moves Luxury Handbags in Mere Minutes." *Business of Fashion*, 4 Mar. 2017, www.businessoffashion.

com/articles/business-blogging/mr-bags-tao-liang-china-bloggerblogger-influencer-sells-thousands-of-luxury-handbags-rewardstyle.

4. "MKI and DocuSign Announce Japan Distributor Agreement for Digital Transaction Service." 6 Sept. 2016, www.docusignDocuSign.com/press-releases/mki-and-docusignDocuSign-announce-japanJapan-distributordistributor-agreement-for-digital-transaction-service.

5. O'Malley, Katie. "Xenia Tchoumi: How Blogging Turned into an Incredible Career as a Model and Entrepreneur." *ELLE UK*, 9 Mar. 2017, www.elleuk.com/life-and-culture/culture/longform/a34418/xenia-tchoumi-female-empowerment/.

6. *HYPR Influencer Database*, 8 Oct. 2017, "Xenia Tchoumi".

7. whinderssonnunes, https://www.youtube.com/user/whinderssonnunes.

8. Wang, Lianzhang. "Dad Launches Startup That Puts Toxic Kids' Products to the Test." *Sixth Tone*, 11 Oct. 2016, www.sixthtone.com/news/1416/dad-launches-startup-that-puts-toxic-kids-products-to-the-test.

9. EeOneGuy, https://www.youtube.com/user/EeOneGuyEeOneGuy.

10. Interviewee requested that both she and her company be anonymized.

5

Business to Consumer (B2C) Influencer Marketing Landscape

It is June 2016, and Isaac Larian, founder and chief executive of MGA Entertainment, the largest privately owned toymaker in the USA, cannot sleep.[1] Ever since his firm's first hit (a talking doll called "Singing Bouncy Baby") was released in 1997, the 62-year-old Iranian-American has become accustomed to sleepless nights. While his company went on to produce several best-selling products, including the Bratz doll franchise, and to acquire the highly profitable Little Tikes children's brand in 2006, success appears to have plateaued.

Ten years have passed since the Little Tikes acquisition and Larian is lying in bed, staring at the ceiling, and restlessly thinking about the next big thing. To seek out inspiration, he turns his computer on and opens YouTube. A few keyword combinations later, he stumbles upon an online subculture of "toy unboxers." This sparks a new obsession and lays the foundation for MGA's next big product.

"Toy unboxing" may be one of the most popular content categories on YouTube you have never heard of. It is as straightforward as the name suggests—YouTubers film videos that show themselves opening brand-new toy boxes. They break down each step of the toy-opening process so viewers can see exactly how everything is packaged, and what all the toy components look like. Kids around the world love watching them.

The most successful "unboxers" draw in viewers through their own excitement, and their contagious enthusiasm gets kids to watch the same videos again and again...and again. According to eMarketer, children ages 11 and younger are one of the fastest growing audience segments for online video

© The Author(s) 2018
J. Backaler, *Digital Influence*,
https://doi.org/10.1007/978-3-319-78396-3_5

(they are growing almost four times as quickly as viewers between the ages of 18 and 24[2]). This means toymakers are shifting resources away from traditional, expensive television commercials and investing heavily in developing influencer relationships, building brand awareness, and driving sales through the "unboxers" and other toy review YouTubers.

Isaac Larian does not know about all of this, but he does know he cannot stop watching these videos, and he is pretty certain most people, especially the children who play with MGA's toys, feel the same way. The next morning, he rushes into MGA's office to share what he learned during his late-night "market research session" and challenges his team to develop the ultimate unboxing toy.

It takes five months, but his team delivers, creating the L.O.L. Surprise, a plastic sphere wrapped in stickers. When a child removes each sticker, they pull out a plastic doll accessory (a purse, mobile phone, jewelry, etc.). After all the stickers have been removed, they open the sphere to reveal two more small plastic bags containing another accessory and the doll itself. Essentially, the effect was a toy that could be unboxed multiple times, and YouTubers ate it up. The product sold out in less than two weeks without any television ads and set the stage for an even more successful product launch, the L.O.L. Big Surprise, a giant sphere with 50 tiny items to unwrap in 2017. Kids went crazy for the Big Surprise, causing it to quickly sell out at most major retailers—again without spending any budget on television advertising.[3]

It took Larian recognizing the way his target consumers purchased MGA's products had changed to understand the value toy Category Influencers on YouTube could have on his business. His company previously spent millions of dollars on television ads to air during children's programming, hoping for the perfect timing when a kid watches the ad and turns to Mom or Dad and says, "I want that!" By shifting focus to where its consumers spend time learning about new toys today (YouTube content creators), and tapping into a powerful trend in the online community (unboxing), MGA developed a line of products and a marketing approach that is fueling the firm's next phase of growth.

B2C Influencer Marketing: Where We Are Today

The toy industry is just one of many consumer-facing industries that are reengineering their traditional advertising models to attract consumers in the digital age, and influencer marketing is a key component of their plans. Certain verticals like fashion, beauty, toys, consumer electronics, alcohol, fitness, and wellness—verticals that have been early adopters—have done very well with influencer marketing. Gil Eyal, chief executive officer of HYPR, an influencer marketing technology firm, explains, "We're seeing companies in these industries shift the majority of their advertising to influencer marketing – companies like Mattel are spending an enormous amount of time and energy to engage Micro-Influencers to get them to talk about their products."

Beauty companies like L'Oreal are long-accustomed to working with influencers; now instead of celebrities, they are more focused on beauty Category and Micro-Influencers. Emily Rubin,[4] a senior marketing executive for a global beauty brand, explains, "If you're in the beauty industry, you need influencers, because it is all about how you look. Back in the 80's, 90's and early 2000's, it was very much driven by celebrities. If you just look at the pages in a beauty magazine it was very much about that authoritarian top-down voice."

But, social media platforms have caused a major shift in their industry. "At this stage, beauty is much more similar to a standard fast-moving consumer goods company," Rubin says. "You want to know what people are talking about with regard to the product, how they are seeing the product today, and how it fits into their lifestyle, and therefore what some keywords are that they are using to learn about the product?" This information is forcing a lot of companies to change their models of how they use social listening to stay on top of the trends and be on trend, and influencers are a major part of that because traditional celebrities are not as effective as they once were.

According to Matt Britton, chief executive officer of the influencer marketing firm CrowdTap, the influencer market for consumer companies still needs time to mature. "Right now, you have a lot of generalists – people who just have a lot of followers and are not particularly influential for anything. In the fashion and lifestyle space you have Instagram models – they're not really models, they're not really curators of any type of fashion or apparel. If they're not influential for one thing, then they're probably not going to be able to move the needle for your business."

Britton feels it is important for brands to take a step back and understand the topic matter in which somebody is influential. "In real life, if you know somebody who is a foodie, you are going to ask them where you should eat – somebody that's well-traveled, you're going to ask them where to go on vacation, etc. That's how people act in the social Web as well. I think that basic principle has kind of gotten lost with all of these numbers and that's where brands can get hurt – they chase the numbers and not really the qualitative side for what somebody is truly influential for."

The timing is still right for B2C brands across verticals to begin reconsidering how they spend marketing budgets, since traditional marketing channels designed to reach consumers are no longer as effective as they once were. Ten years ago, on average, a person needed four exposures to an ad before they took action—meaning if they had an interest within four exposures, the average person would follow through, get the necessary information, and make a purchase. Today that number has quadrupled. It now takes 16 exposures before the average person takes some form of action.[5] According to research conducted by Collective Bias,[6] a consumer and retail-focused influencer agency, influencer content is viewed seven times longer than an average digital ad (19s for a display ad versus two minutes 18s for influencer content).

7.4 X's

Influencer content is viewed 7 times longer than the average digital ad, according to *Collective Bias'* Influencer Trends Report

---

This is an important takeaway for consumer companies, with influencer marketing presenting a better alternative than spending big budget dollars to get their brand in front of consumers, without much guarantee they will see results. The same research indicates that viewership of influencer content mirrors product category sales by season and reflects actual sales performance in the marketplace for consumer products, further demonstrating the effectiveness of influencer marketing for B2C companies. Influencer marketing allows consumer brands to connect with consumers in a more organic way, rather than adopting the disruptive "interruption marketing" approach introduced in Chapter 2.

Interested in B2C Influencers? Don't Forget These Considerations

Despite the benefits influencer marketing can offer B2C companies, it can be challenging to execute. It is a lot easier to pay for television commercials, ad campaigns on Facebook or Google AdWords than to work with influencers. There are challenges brands need to be aware of at each step of process, from identifying the right influencers, developing an effective influencer outreach strategy, collaborating with influencers, and measuring the return on investment (ROI) (not to mention potential brand risks to watch out for). All of these topics are covered later in this book and are relevant to both B2C and B2B brands. What follows are a series of considerations specific to B2C brands to be aware of when deciding on their influencer marketing approach.

When Influencers Become Brand Competitors

Chiara Ferragni, an Italian fashion Category Influencer, started her blog *The Blonde Salad* in 2009 as a way to share her unique style and stories from her personal life with readers from around the world. "People have always been so obsessed with my clothes and accessories, and they wanted to know where they're from, so that's where *The Blonde Salad* started,[7]" she says. She began by writing about other brands, and as her influence grew over time, she started working with shoe brands like Steve Madden and Superga.

What these brands did not realize is just how much Ferragni loves shoes, and by the time she reached one million followers on Instagram in 2013, she launched her own brand of footwear. Her "Chiara Ferragni" branded shoes are now sold in more than 300 stores worldwide, including two of her own retail locations in Milan and Shanghai, with plans to open more retail outlets around the globe.

Ferragni's ambitions to start her own fashion brand should not come as a surprise. Think back to Chapter 4 and the story about Melilim Fu, the Chinese cosmetics Category Influencer whose ultimate goal is to launch her own line of branded products. Another Chinese influencer focused on fashion, Zhang Dayi, runs a successful e-commerce store featuring her own designed and manufactured products and which brought in $46 million in 2016.[8]

"When you think about it, an influencer is a brand – a personal brand. Therefore, when a corporate brand wants to collaborate with an influencer, it's really two brands trying to come together," explains Taryn Southern, a prominent YouTube personality who also advises influencers on monetization strategies. "Those two brands could have very different values or ways of engaging their target audience, which can complicate things, so going at it alone when the timing's right is a great option for influencers."

"An influencer is a brand – a personal brand – therefore, when a corporate brand wants to collaborate with an influencer, it's really two brands trying to come together."

-Taryn Southern (@TarynSouthern) YouTube Influencer & Digital Strategist

DigitalInfluenceBook.com   SHARE THIS

Fig. 5.1 Taryn Southern

As influencers gain more and more influence and learn about the likes and dislikes of their loyal followings, it is natural for them to choose to develop their own branded product lines. There are already plenty of cases studies out there that provide influencers with a blueprint to forge their own path rather than rely on a corporate brand's existing platform. Therefore, B2C brands should be cautious when collaborating with influencers, and always know that in the back of the influencer's mind, they may be thinking about how they could do it better with their own brand (Fig. 5.1).

When Influencers Become Co-creators

Not every influencer aspires to develop their own company, however. For example, operating a cosmetics or clothing e-commerce store is another full-time job that comes with all sorts of headaches that many influencers may not want to deal with. Manufacturing, managing inventory, customer service—these are all outside the core skills required to be a successful content creator. Influencers may simply prefer to continue to focus on their passion, creating engaging content for their loyal audiences. This is where a longer-term co-creation partnership with a brand that an influencer genuinely enjoys could be a good fit.

To date, co-developed and co-branded influencer collaborations are most prevalent in the fashion and beauty industries, but the strategy is certainly applicable for brands in other consumer-facing verticals. In 2015, Nordstrom, the American luxury retailer, teamed up with Emily Schuman, the face behind the popular blog *Cupcakes and Cashmere*. Over the course of 18 months, Schuman and Nordstrom's design team developed a new clothing collection that went on to be sold in Nordstrom outlets under the Cupcakes and Cashmere brand. Nordstrom leveraged Schuman's understanding of her audience to develop a unique product line that fit their needs, and the Cupcakes and Cashmere fashion brand lives on to this day.

According to Beca Alexander, founder and president of the influencer marketing agency Socialyte, brands are starting to become more willing to enter into these longer-term partnerships with influencers. "There's currently a trend in the market where brands that have identified the right influencers that they want to continue to work with in a larger capacity are signing longer-term ambassador contracts that don't allow them to work with competitors in their category," she says. "Instead of saying, 'I don't want to work with influencers who have worked with my competitors,' they're saying, 'I want to work with this person, and I don't want them to work with any of my competitors for a year, so I'm willing to sign a contract with them for a year or multiple years and pay them for exclusivity.'"

This type of long-term-oriented brand-influencer partnership can be extremely beneficial for both sides. The brand benefits from gaining valuable customer insights to fuel product development, and an active promotional channel to reach their target audience. Meanwhile, the influencer monetizes their expertise and online following without having to take on a new set of business duties. Of course, for the influencer they could make more money going at it on their own, but the value-add of an established brand that can handle the business side of things may make the smaller paycheck worth it. Finally, for brands that consider this co-development, co-branded strategy, be sure to read Chapter 10 carefully to fully understand the risks of having close ties with an influencer's brand.

When Influencers Become Distribution Channels

Another way consumer brands can think about influencers is to approach them as distribution partners, especially when it comes to doing business in international markets. Since distribution partners understand the local consumer landscape, they are often the best option for companies when they first consider expanding into a new country. The partner relationship helps get products into the market faster and avoids expensive investments in local offices until the market is proven. A 2016 Harvard Business Review article[9] featuring highlights from a global distribution study conducted by Frontier Strategy Group, which found that third-party distribution partners account for anywhere between 41 and 72% of corporate revenues in emerging markets—that means the right local influencer could help add meaningful contributions to a company's revenue figures.

How does this work in practice, though? Think back to the ZooEnglish. com example in Chapter 4. The consumer brand sold its software product licenses directly through the WeChat accounts of early childhood education influencers to generate sales in China. The influencers themselves had no intention of building their own English-learning software, nor did they wish to collaborate with ZooEnglish to co-brand a new solution for their audience. All the influencers wanted to do was to identify quality early education products from the USA and Europe that their audience of engaged Chinese parents would find value in and strike distribution sales agreements with the foreign brands to profit as an intermediary (Fig. 5.2).

B2C brands should be on the lookout for influencers at home and overseas who have the attention of their target consumers. With the right approach, brands can strike mutually beneficial deals with these influencers to sell products via the influencers' social media channels, or incorporate

Fig. 5.2   HBR-Frontier Strategy Group distribution statistic

their products into an influencer's existing online storefront. Regardless, given the power influencers have on consumer purchase decisions around the world, brands would be wise to start thinking more creatively about who they approach when it comes time to strike a new distribution partner deal.

Solely a Distribution Channel? What About All That Content?

While influencers can help serve as distributors to sell brands' actual products to their audience, brands most often turn to influencers as content distribution channels to get targeted messages (like product promotions or new launches) in front of influencers' audiences. Today influencers are the new gatekeepers for brands who want to reach targeted consumer audiences, but influencers are much more than simply distribution channels. One of the greatest points of frustration among influencers and industry insiders interviewed for this book is how often the brands they work with do not realize how much effort goes into creating the content that goes in front of their audience each and every day. A common feeling, often expressed in different ways, boils down to they view us just as a way to increase reach, but what they do not seem to value is all the work their expensive creative agency would typically do, I do as well. Content (blog posts, photos, videos, podcasts, etc.) does not simply appear out of nowhere, it takes time, energy, and often upfront financial investment as well.

Reb Carlson, associate account director at the digital agency Wunderman, sees the same on the agency side. "One thing that I think a lot of brands don't appreciate is how much time original content takes. Even a YouTube video will take a full day of filming." She has the following advice for brands: "Brands need to understand Influencers are not just posting to the audience, there's a lot of work that goes in behind the scenes... When I first got into influencer marketing it was more about brokering 'influential content creators' – same as commissioning art or content from someone – there was a lot of respect for the influencer's deliverable. For some reason, social media content tends to be viewed as easier to produce, and therefore the brand is only paying for reach, but not realizing what it takes to produce the creative assets."

For B2C brands who want to tap into Influencers' existing relationship to their online audiences, they would be wise to emphasize your desire to collaborate as much for their reach as the content they produce. By at least acknowledging the Influencers' efforts on the content production side of the relationship, the brand can separate itself from the countless others who reach out to simply get in front of their followers.

Compensating B2C Influencers: Can We Pay You in Jewelry?

When Los Angeles-based "mommy bloggers" Erin Ziering and Veena Crownholm teamed up at the end of 2016 to launch their Millennial Mamas brand, they quickly gained traction, and brands targeting influential millennial mothers came across their website and social channels. The duo uses their marketing savvy and understanding of the aspirations of other millennial mothers to develop family-friendly, "Instagrammable" experiences for brands, including Sony Pictures, Unilever's Breyers Ice Cream, and The Grove retail complex in LA (Fig. 5.3).

Crownholm explains, "There are definitely times where we choose strategic partnerships with certain brands that can get us in front of the right audiences and add value to both the Millennial Mamas community as well as the other brands we partner with on an ongoing basis. These brand-sponsored events give us exposure to a large group of millennial mothers who can benefit from our content. Ultimately, being able to convert these brands'

Fig. 5.3 Erin Ziering and Veena Crownholm, *The Millennial Mamas*

fans into our own following will lead to more lucrative opportunities for us in the future."

For influencers, especially Micro-Influencer or emerging Category Influencers, consumer brands do not always need to pay cash to work with them. The brand cache of being able to say, "I collaborated with Brand X" or an offer to send free product samples may be sufficient. However, as Influencers' audiences grow, and they have more negotiating leverage, there is an expectation across most consumer verticals that brands will pay influencers to collaborate. Not only is there an expectation, but if a brand reaches out to an established Category Influencer or Celebrity Influencer for the first time proposing collaboration in exchange for free products, they will likely either be ignored, or even worse, publicly shamed on the Influencer's social media platforms.

Part of why payment is so essential as influencers gain experience is they tend to involve more support staff to help manage different aspects of their influencer empire. Take Chiara Ferragni for example. Ferragni has a team of 20 employees at the *Blonde Salad*[10] with titles including office assistant, project manager, e-commerce manager, accounting coordinator, and so on. On the other end of the spectrum, established influencers will have a few support staff like assistants, bookkeepers, or graphic designers. At the very least, they will have an agent to help book paid opportunities on their behalf. What this means for brands is incentivizing through unique experiences or free products are a non-starter because they do not pay the bills (Fig. 5.4).

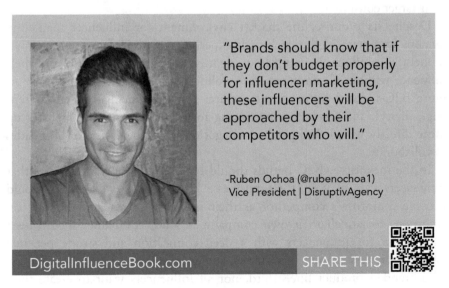

"Brands should know that if they don't budget properly for influencer marketing, these influencers will be approached by their competitors who will."

-Ruben Ochoa (@rubenochoa1)
Vice President | DisruptivAgency

DigitalInfluenceBook.com   SHARE THIS

Fig. 5.4 Ruben Ochoa

Ruben Ochoa, an influencer marketing specialist and vice president at DisruptivAgency, recalls his past experience: "I've had jewelry brands be turned down by influencers with only 50,000 followers because the brand wanted to pay in rings and necklaces. The problem is that products can't be split among influencers, their managers or anyone else involved in the development of their content." He urges brands to realize that "it's a business now, not the wild west and brands should know that if they don't budget properly for influencer marketing, these influencers will be approached by their competitors who will."

Getting B2C Influencer Marketing Right

When MGA Chief Executive Isaac Larian spent the night watching unboxing videos in June 2016, not only did he stumble upon a new marketing channel for MGA's next series of products, he also came to realize that by turning to influencers for guidance on the latest products, his target audience was no longer spending hours in front of the TV watching the traditional ad segments that cost his firm millions of dollars each year. This shift is one impacting brands across consumer verticals such as fashion, beauty, toys, consumer electronics, alcohol, fitness, and wellness. As a result, these industries are reallocating budget dollars to digital, with a heavy emphasis on collaborating with influencers to gain the attention and wallet share of their target consumers.

Despite its promise, the market environment for influencer marketing remains in a state of flux as brands experiment with different engagement models, and influencers struggle to identify the right way to monetize their influence without alienating the very followers that give them authority. As more B2C-focused influencers transition from producing general content (e.g., lifestyle influencers) to narrowing focus on content that matches up more directly with brand product categories, it will be easier for both sides to collaborate.

In the meantime, B2C brands should be aware of the various roles that influencers can play in their business. First, today's collaborator could become tomorrow's competitor as their influence grows and they act on ambitions to launch their own company. Second, companies should keep an eye out for influencers with a deep connection and strong personal brand awareness among their target audience and consider co-developing a co-branded product line. Third, not all influencers want to create their own or collaborate on new products, some simply want to strike a mutually beneficial agreement to distribute products that will add value to their engaged audience.

Remember, when it comes to influencer collaboration, a lot of effort goes on behind the scenes in the content creation process. It is important to recognize and properly incentivize influencers for their efforts both as creators and as distribution channels. Lastly, know when it is appropriate to offer product or experiences in exchange for influencer collaboration, and when it is not. Odds are more established Category Influencers and Celebrity Influencers will only accept cash since they have more fixed costs in running their business, so they need to be able to pay the bills.

While influencer marketing is more established for B2C brands, B2B brands also have much to gain by incorporating influencers into their business strategies. Turn to Chapter 6 for plenty of key takeaways for B2B firms to learn from, including Walter Jennings' story of developing the Chinese technology company, Huawei Technologies, global B2B influencer relations practice.

Marketer's Cheat Sheet

- **An Essential Channel**: Certain B2C verticals, especially toys, fashion, and beauty, are shifting budget dollars to influencer marketing in a big way, as their consumers turn to trusted online voices for guidance on what to buy and the latest trends.
- **Quality Varies**: While there are many B2C influencers who align well to consumer brands and specific product categories, there are still others getting hired who are more generalists and do not produce the same positive results.
- **Influencers are Brands**: Especially in a B2C context, influencers are their own personal brands; therefore, when a corporate brand wants to collaborate with an influencer, it is really two brands trying to come together, and those two brands could have very different values or ways of engaging their target audiences.
- **Influencers as Competitors**: Not every influencer is satisfied making a career out of collaborating with established brands; instead, they prefer to leverage their platform to launch their own companies such as Chiara Ferragni.
- **Influencers as Co-Creators**: Brands can align themselves with influencers who may otherwise want to launch their own brand, but lack the knowledge and resources to do so, and work together to co-create a co-branded product line, similar to how Nordstrom collaborated with Emily Schuman to co-develop the Cupcakes and Cashmere product line.
- **Influencers as Distributors**: Some influencers simply want to produce content and engage an audience with products developed by established brands, like ZooEnglish's Chinese WeChat influencers. B2C brands can tap into the power of these category influencers, especially when approaching new international markets.

- **Manage Expectations Appropriately**: Ensure influencers are compensated both for the content they produce and their promotional channel, and do so in a way that makes sense for their level of influence. Free product offers and trips to exotic locations will only go so far in laying the foundation for successful long-term partnership with an effective influencer.

Notes

1. Stratton, Alexandra. "Toymakers Curry Favor with Precocious YouTube Influencers." *Bloomberg.com*, 18 Oct. 2017, www.bloomberg.com/news/articles/2017-10-18/toymakers-curry-favor-with-precocious-youtube-influencers.
2. Pierson, David. "My Kids Don't Have a YouTube Channel—But They Pretend They Do." *Los Angeles Times*, 27 June 2016, www.latimes.com/business/technology/la-fi-youtube-kids-20160627-snap-story.html.
3. Stratton, Alexandra. "Toymakers Curry Favor with Precocious YouTube Influencers." *Bloomberg.com*, 18 Oct. 2017, www.bloomberg.com/news/articles/2017-10-18/toymakers-curry-favor-with-precocious-youtube-influencers.
4. Illustrative name, actual interviewee wished to remain anonymous.
5. Song, Kerry. "What Is Influencer Marketing?" *Tonyrobbins.com*, www.tonyrobbins.com/career-business/what-is-influencer-marketing/.
6. Collective Bias. *Influencer Trends Report*, May 2017, https://collectivebias.com/blog/2017/05/influencer-trends-report-a-spotlight-on-the-power-of-influence/.
7. O'Connor, Clare, et al. "Meet Chiara Ferragni, The Fashion Influencer Behind The Blonde Salad." *Forbes*, 2017, www.forbes.com/video/5588620538001/.
8. Pan, Yiling. "Top Web Celebrity Zhang Dayi Reveals the Key to Her Business Success." *JingDaily*, 19 July 2017, https://jingdaily.com/uncovering-business-secrets-chinas-top-web-celebrity-zhang-dayi/.
9. Brier, Ryan. "The Sales Role Multinationals Need in Emerging Markets." *Harvard Business Review*, 31 Oct. 2016, https://hbr.org/2016/10/the-sales-role-multinationals-need-in-emerging-markets.
10. The Blonde Salad. *Team Page*, https://www.theblondesalad.com/tbs-crew.

6

Business to Business (B2B) Influencer Marketing Landscape

Walter Jennings, an American expat based in Hong Kong, has his work cut out for him. After a successful career spanning nearly two decades of working for established Western multinational companies in various senior global communications roles, he decided to take on a new challenge and join Huawei Technologies, a fast-growing Chinese technology company. It is not uncommon for foreigners to work for Chinese companies, but more often than not, at the management level their role is all too often a "foreign face," a figurehead to help "internationalize" the brand since these firms are household names in China, but are relative unknown overseas.

Jennings' role is unique, however. Foreign hires at other Chinese companies tend to work out of remote offices in the USA or Europe and have very little decision-making authority or visibility into their management team's true ambitions; Jennings is based in the firm's Chinese headquarters alongside a predominantly local senior management team. This gives him a lot more power, visibility, and internal influence. Also, unlike most Chinese companies, his new employer is already VERY global, operating in more than 170 countries around the world.

The challenge for Jennings is that Huawei has a long history of being misunderstood. Everything from the brand name's pronunciation (pronounced HWA-way) to its ownership structure has proven a challenge for foreign media and industry peers. This is mainly because early on in its development, the company's primary focus was on innovation, sales, and international expansion. Very little attention was placed on business areas like external communications or corporate reputation management, which left

© The Author(s) 2018
J. Backaler, *Digital Influence*,
https://doi.org/10.1007/978-3-319-78396-3_6

outsiders to draw their own conclusions about the firm's motivations based on what limited information existed in the public domain.

This all came to a head in 2012, when Huawei was summoned to a special hearing in front of the US House Intelligence Committee. Huawei, and its Chinese competitor ZTE, failed to demonstrate to the special US committee that their products were safe for use by American companies and the US Government. In an official statement, Huawei Founder and CEO Ren Zhengfei explains, "We have never sold any key equipment to major U.S. carriers, nor have we sold any equipment to any US government agency. Huawei has no connection to the cyber security issues the US has encountered in the past, current and future."[1] But his statements to the press came too late and were not enough to reshape the firm's reputation and to prevent the special committee from concluding Huawei's telecommunication equipment should not be used by the US Government or American companies.

Experiences like these helped Huawei's leadership come to realize that reputation management is not something that happens organically. They needed to take a more proactive approach and focus attention and resources on "corporate reputation management," which is where Jennings comes in. His official job title is "VP of Corporate Communications," but the vast majority of his time and energy is dedicated to building and managing Huawei's KOL (key opinion leaders) program, a community of Category Influencers who produce content for their own audiences on topics relevant to Huawei's business, including Internet of Things, virtual reality, autonomous driving and next-generation telecommunications technologies. At the time of the writing of this book, Huawei's KOL program includes more than 100 influencers from a diverse range of markets: Australia, Malaysia, India, Pakistan, Saudi Arabia, UAE, Nigeria, South Africa, Poland, Russia, Turkey, France, Ireland, Canada, USA, and Mexico.

The basis of KOL-engagement is exclusive experiences with the Huawei brand at its international conferences. Jennings invites a group of anywhere between 10 and 20 KOLs to each event, provides them with business-class airfare, five-star hotel accommodations, and a front row seat to the action. Once at the conference, Jennings personalizes the experience by facilitating interactions with new technologies and access to Huawei executives, which in turn helps the KOLs generate new content and inspiration to engage their own followings (Fig. 6.1).

It took more than two years of trial and error to reach this point. When he started, Jennings did not have a playbook to identify the right influencers, build influencer relationships, or work with influencers—he experimented with different approaches, technology solutions, and agencies, which led to some inevitable bumps along the way.

Fig. 6.1 Huawei KOL Dez Blanchfield experiences the latest holographic communications technology

Shel Israel, a Category Influencer focused on Augmented Reality (AR), attended the first Huawei KOL gathering at the Mobile World Congress in Barcelona in February 2016. "I was almost totally immersed in AR and VR at the time. Huawei's PR representative promised me there would be much to see," he says. "Within an hour of the first session I was told there was little or no AR involved in what I would see or hear about. This made me suspicious, and left me wondering why I had been invited at all." Following the event, Israel made sure to let Jennings know about his disappointment.

A few weeks later, Israel received a brief email from Jennings with the subject line, "I fired our public relations agency…" It took hiring a new agency and several months of relationship management, but Israel went on to attend future Huawei KOL conferences: "It's been fascinating to watch their program's transformation over the last two years," he says. "It's now much more mature, with an engaged network of high-quality KOLs that are helping to build awareness about the Huawei brand among relevant audiences" (Fig. 6.2).

Moving forward, the challenge for Huawei is scale. Jennings is the primary point of contact for every single KOL in the program, and his only support staff is an external public relations agency that provides project-based support when influencers travel to Huawei's international conferences. He even personally manages a private Facebook group for the KOLs, extending their offline experience at Huawei's conferences into a virtual, collaborative community with participants from around the world.

Fig. 6.2 Jennings alongside Huawei KOLs at its 2017 Mobile Broadband Forum in London[2]

In interviews with Huawei KOLs for this book, nearly all of them point to Jennings as the reason why they enjoy participating in the program. "Walter creates amazing experiences for the KOLs," explains Victoria Taylor, founder and chief digital strategist at Untwisted Media (who also serves as one of Huawei's UK-based KOLs). "These events give us a unique opportunity to interact with the latest technologies and connect with fascinating peers in the KOL group which helps inspire new innovation and thought leadership. He really personalizes the experience, which is part of what makes the program so unique."

In a blog post following Israel's second Huawei experience, he also praises Jennings' efforts: "Walter is important to this post because he is the glue that binds together the loosely structured network called KOL. I don't know how the concept actually got started, but to each of us in KOL, Walter feels like a personal friend and demonstrates a profound sensitivity to what each of us is attempting to accomplish.[3]"

B2B Influencer Marketing: Where We Are Today

Jennings' grassroots approach to building Huawei's KOL program is not as unique as one may think. While B2B companies have a long history of working with journalists and industry analysts, broader Category Influencer engagement is relatively new to even the world's largest, most-respected B2B brands. As a result, all B2B brands appear to be experimenting with the right strategy to work with influencers in a B2B context to maximize results. While there are certainly aspects of B2C influencer engagement that B2B practitioners can learn from, it is far from a "copy and paste" blueprint— B2B requires a specialized approach tailored to the realities of the companies' business models and the different types of influencers they work with.

For starters, terminology and engagement models vary significantly between B2B and B2C brands. In the B2B space, "influencer marketing" is most often classified under the title "influencer relations." For a long time, B2B companies felt as though because they have customers and not consumers (different ecosystems of buyers), they thought influencer marketing really had no bearing on what they did. They thought it was really only something for consumer product companies—what would work for L'Oreal and its cosmetics wouldn't make sense for IBM and its Watson artificial intelligence solutions, for example. Now, we are starting to see a lot of B2B companies understand that there is a role for influencer marketing—they just call it influencer relations instead. To understand the difference between "influencer marketing" and "influencer relations," consider Company X, a large technology firm with two divisions: the Consumer Group and the Enterprise Group.

The Consumer Group markets to individuals for the purpose of selling products like tablets and smartphones. Its audience is made up of individual "consumers" who make purchase decisions relatively quickly. (Do I need a new phone? Yes. Do I believe this is the best phone for me? Yes; I'll buy it.) Through "influencer marketing," Company X can help drive sales and awareness of its consumer products in the same way as the B2C companies do in Chapter 5. Oftentimes, these influencer relationships are one-offs and are only tied to specific initiatives. Once the particular initiative concludes, the influencers and Company X's representatives go their separate ways. However, Company X is not purely a consumer-facing brand so it also has a separate function that works with a different set of B2B influencers on an ongoing basis.

The Enterprise Group markets the same products, but to corporate customers (B2B) instead of individual consumers (B2C). Rather than aiming to get an individual to purchase a smartphone for $1000, enterprise needs to convince its corporate buyer to purchase several millions of dollars-worth of smartphones and tablets to outfit all of their employees worldwide. Unlike a short-term, point-in-time B2C sale, B2B sales cycles may last one to two years before corporate customers are able to switch vendors or purchase a new solution. Influencer relations departments engage the key opinion leaders that their end-customer turn to for advice about B2B solutions. By building long-term, trusted relationships with KOLs (aka Category Influencers), B2B companies are able to drive awareness and help educate potential customers over the course of extended periods of time, so that when the customer is finally able and ready to make a decision 12–18 months later, Company X's product becomes the natural choice (Fig. 6.3).

Tim Crawford, CIO strategic advisor at AVOA, a US-based business consultancy, is also a Category Influencer on topics relevant to chief information officers. He recalls a past experience where he influenced a series of B2B deals: "I can think of one occasion where I had a relationship with executives from two B2B companies and I introduced them to each other, given my knowledge of one party's needs, and the other party's solutions. An initial deal six months later, worth a few hundred thousand dollars between the two firms, evolved into a multi-year relationship amounting to over hundreds of millions of dollars exchanged between their companies. It all started with my introduction."

Perhaps social media marketing specialist Mark Schaefer puts it best in his white paper, *The Rise of Influencer Marketing in B2B Technology*. Co-authored with the influencer software firm Traackr, it explains how influencers have become a critical vehicle for B2B buyers:

"In B2B, we're moving away from Brands and towards People – customers are more attached to industry influencers than the brands that employ them."

-Mary Shea (@sheaforr)
Principal Analyst, B2B Marketing
FORRESTER RESEARCH

SHARE THIS

Fig. 6.3 Mary Shea

For today's new breed of B2B buyers, companies may be seen as cold and detached, but their favorite online experts are trusted friends. We'll ignore an ad, but we'll subscribe to all of the content from our favorite influencer.

On the surface, the difference between "sales" and influencer marketing is reach and trust. Even after a decade of work, a B2B salesperson may have a limited file of connections but an influencer may own the hearts of thousands or even millions of fans. A company representative has an agenda, so there is natural mistrust. An influencer, in theory, is a passionate, honest expert.[4]

These words hold true for any B2B company, not just technology companies—management consultancies, financial services firms, manufacturers [*insert B2B industry name here*], all stand to benefit from building a thoughtful influencer engagement strategy. The fundamental way corporate buyers purchase high-cost products and services has changed dramatically. The world where relentless salespeople "dial for dollars" and barrage prospects with unsolicited email blasts no longer yields positive results.

As Schaefer rightly points out, today when people have a need, they don't wait for a sales call. They proactively seek out trusted advisors in the space who can help educate and inform their purchasing decision. It is critical for B2B brands to understand who those trusted Category Influencers are that their own customers and prospective customers turn to for advice and then develop direct brand–influencer relationships to increase the odds their company's solution is at the top of the list when the opportune moment for a recommendation arises. The next section introduces a few things for companies to keep in mind when approaching these B2B influencers.

Interested in B2B Influencers? Don't Forget These Considerations

While influencers have a role to play for B2C (see Chapter 5 for more detail) and B2B brands, the brand–influencer relationship varies significantly between the two. For example, with top influencers, B2C companies almost always have to pay to incentivize the influencer to collaborate, and there's also risk that B2C influencers could become future competitors as they grow their following and develop their own brands to market. It's a very different situation in the B2B influencer world, and the following are a few key considerations to keep in mind when building a B2B influencer strategy.

A Day in the Life of a Busy "B"-2B Influencer

Lee Odden is CEO of TopRank Marketing, a digital marketing firm based in Minneapolis, Minnesota. In addition to his full-time job, he is also consistently voted as a leading B2B influencer for companies focused on the topic of content marketing. He wears multiple hats and his "influencer" hat is constantly at odds with his "full-time CEO" hat.

Each decision to collaborate with other brands needs to be weighed against the pros and cons of how his participation will impact TopRank. Influencer opportunities, such as international conference keynotes, webinar collaborations, and white paper contributions, collide with upcoming client deadlines, internal commitments, and unpredictable personnel issues. He did not set out to consciously "become an influencer." However, through years of consistently publishing high-quality material about content marketing, he naturally grew his audience, which caught and continues to catch the attention of brands, conference organizers, and media alike.

Herein lies the conundrum that many B2B influencers face: Their number one priority is a full-time job (either their own company or someone else's company). In theory, influencer-related activities take valuable time and attention away from their day-to-day professional responsibilities, but in reality, the additional visibility they garner from speaking at a conference or participating in a podcast interview builds awareness of both their personal brand and their employer's brand, which ultimately leads to new business opportunities for the company. The ultimate conundrum is how to strike the right balance: It is all too easy to say "yes" to every influencer-related request, but if the Influencer agrees to participate in every opportunity, then they will be unable to fulfill the obligations of their day-to-day role.

Unlike in the B2C space, where many individuals operate as full-time influencers, B2B influencers are most often like Odden—they either own their own company or work for someone else's company, which leads to a variety of interesting issues unique to working with B2B influencers as compared to their B2C counterparts.

For B2B Influencers, Objectivity Is Everything

Objectivity is the lifeblood of everything B2B influencers do, even more so than with B2C influencers. Jill Rowley, a prominent evangelist for "social selling," explains how B2B influencers need to be cautious when choosing which companies to collaborate with: "As a sales professional, me being connected to and engaged with the influencers of my buyers makes me more essential. When I first went [out] on my own, I took every opportunity to get the word out, but now I have to be a lot more careful."

She recalls an instance during which she participated in a video interview with LeadIQ, a sales enablement software company. Afterward, one of her contacts reached out to learn more about the product that was discussed. "I didn't talk about [LeadIQ's] product at all in the video," she says. "But I realized that just by being part of their content marketing strategy, our association could be viewed as an indirect endorsement. I need to be really careful" (Fig. 6.4).

B2B audiences, especially at the executive, decision-maker level, tend to be highly educated. They understand the issues within their industry quite clearly, and they are generally aware of the various solution offerings for their role. This means B2B influencers operating in the space need to be extremely cautious, as they are dealing with a very skeptical, discerning audience who will call them out on any white lies or half-truths.

Influencers who shamelessly promote one brand's product will lose their credibility in an instant, so they need to take a more indirect approach when collaborating with brands. "As an influencer, I'm 'unbranded.' I don't just work with IBM exclusive to another brand," explains Tamara McCleary, a B2B influencer among a chief marketing officer audience. "If I only work

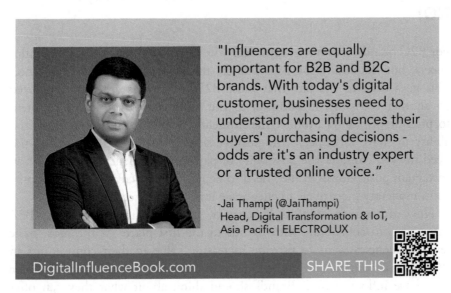

"Influencers are equally important for B2B and B2C brands. With today's digital customer, businesses need to understand who influences their buyers' purchasing decisions - odds are it's an industry expert or a trusted online voice."

-Jai Thampi (@JaiThampi)
Head, Digital Transformation & IoT,
Asia Pacific | ELECTROLUX

DigitalInfluenceBook.com   SHARE THIS

Fig. 6.4 Jai Thampi

with a single brand as an influencer in the B2B space, no one is going to listen to what I say. They know it's ridiculous that one brand would have all the answers. The most influential influencers in the B2B space work with all of the companies in a given category as an unbiased source to talk about not only the solutions, but the problems and challenges too, which ultimately builds trusted relationships through telling the truth. To have influence you must become a trusted leader, mentor, and guide within a given industry. Transparency, and honesty are critical in building influence."

McCleary goes so far as to challenge B2B brands who think they cannot work with her because she has already collaborated with one of their competitors: "That's exactly why you need to work with me—you should want to work with influencers who are working with your competition, because that's a signal that the influencers have an intimate understanding of your industry and can speak objectively about the current landscape."

Before brands even think about reaching out to a B2B influencer, they should take a moment to consider if the request could be perceived as encroaching on the influencer's objectivity. If the answer is "yes," then it is probably worth thinking of a different form of collaboration to propose instead (see Chapter 8 for more advice on developing an effective influencer outreach strategy).

To Pay or Not to Pay? The Answer Is Always NO!

For B2C brands, it is not always clear whether or not they should pay influencers, but thankfully for B2B brands, the answer is almost always "no." This goes back to the fact that most B2B influencers are fully employed by another company, in addition to the fact that if they accept payment to promote a brand's product the influencer will lose their objectivity and ability to "have influence." The positive side of this is, B2B brands do not need to even consider the possibility of paying an influencer $100K+ for an Instagram post. However, they need to be much more creative in developing "non-monetary incentives" to encourage influencers to collaborate.

How can brands do this? It starts by understanding what drives B2B influencers to collaborate with brands in the first place. At the most basic level, influencers want to provide increasing value to their community, in the form of content and engagement, and to grow the number of people they have influence over. Brands should think about what they can offer that could help the influencer engage his or her existing audience and aid in broadening awareness of the influencer among relevant audiences to increase his or her following (Fig. 6.5).

B2B influencers commonly collaborate with brands by participating in events, leading webinars, co-creating content in the forms of blogs, podcasts, and white papers—the ultimate form of collaboration is only limited by the brand's creativity. Conferences are a common way to build an initial relationship with influencers, because they present a chance for the brand to add value to the influencer and all meet in person where they can establish rapport and have informal discussions about potential additional areas to collaborate that may be more time intensive for the influencer. Some other examples outside of Huawei's conference-led engagement strategy include efforts by American software companies Adobe and Teradata.

Adobe engages a select group of influencers who become part of their Adobe Insiders program. Each year, the group of influencers are invited to the company's flagship event, Adobe Summit. While at the event, "Adobe Insiders" share their experience via social media and impart key takeaways with their online audiences. Teradata holds an annual Influencer Summit event, for which they invite a much larger group of influencers to experience their latest technology. Senior executives introduce new innovations, and influencers have an opportunity to ask questions to help understand the firm's strategy and what products will likely be released over the next several months and years.

"B2B influencers are not just industry bloggers and analysts. CMOs are starting to take a much more public stance to leverage their personal brands to promote their companies."

-Kimberly Whitler (@KimWhitler)
Marketing Professor
DARDEN BUSINESS SCHOOL

DigitalInfluenceBook.com   SHARE THIS

Fig. 6.5 Kimberly Whitler

B2B influencers participate in unpaid brand collaborations for a variety of reasons. For B2B influencers who fall into the "emerging Category Influencer" bucket, brand cache can go a long way. For example, being able to identify as an official brand influencer, the likes of Huawei KOL, IBM Futurist, Oracle Social Champion, and Adobe Insiders all help add to their credibility in the marketplace and among their audience. The opportunity to speak at conferences and participate in other forms of content collaboration can lead to bigger audiences and new content for them to engage with.

Other more established influencers appreciate the dedicated time they get to produce content on the way to and from brand conferences. Dez Blanchfield, an Australia-based digital transformation-focused influencer who works with a wide range of international companies, enjoys the high-productivity output he can get on a long-haul business-class flight. "When I'm in my business class pod, I have zero distractions and can get weeks-worth of work done as a result of it."

Peter Shankman, an award-winning B2B author and influencer, appreciates the uninterrupted time as well. "I've written three of my books entirely while flying on planes. The level of focus when you're in the air away from all distractions is second to none."

Even though B2B influencers can get a lot of value out of collaborating with brands on an unpaid basis, it does not mean that there is never money exchanged between the two sides. In the world of B2B, payment is much more indirect than the overt transactional approach adopted by B2C brands.

Well…Sometimes It's OK to Pay

Let's go back to the statement in the preceding section: "At the most basic level, influencers want to provide increasing value to their community, in the form of content and engagement, and to grow the number of people they have influence over." While this statement is certainly true, there is one caveat—B2B influencers still need to make money. It is fun to fly around the world to attend conferences and exchange ideas with industry insiders, but these experiences do not help pay the bills.

This is where a delicate grey area exists between brand and influencer in the B2B space. Brands understand that influencers ultimately want to be paid, but they also recognize that influencers need be extremely cautious of what they accept payment for in order to maintain their objectivity in the marketplace. The following is a high-level overview of examples in the world of B2B influencers, demonstrating when it is ok to pay and when it is not:

✓ *OK to Pay a B2B Influencer*

✓ Travel expenses for attending a conference
✓ Keynote speech
✓ Workshop facilitation
✓ Webinar production
✓ White paper development

✗ *NOT OK to Pay a B2B Influencer*

✗ Spa services at the hotel while traveling for a conference
✗ Promoted tweets
✗ Promoted blog posts
✗ Promoted articles
✗ Promoted videos
✗ Any form of paid promotion that undermines their objectivity

Outside of direct payment for their time, unpaid collaboration helps make the influencer "known" among various important management stakeholders in the organization. As more people learn about the influencer and their subject-matter expertise, over time this will lead to paid engagements. For example, when a need for consulting services emerges, or when there is an opening to have a workshop presentation at an annual sales conference. These types of consulting opportunities are when it can become very lucrative for influencers to have deep relationships with B2B brands. The key is ensuring that once an influencer agrees to be paid for these types of services, externally they need to continue to treat that brand's solutions just as objectively as they would one of their competitors.

B2B Influencers vs. Analysts vs. Journalists

B2B firms have established practices for working with the media and industry analysts. But as Carter Hostelley, CEO of Leadtail, a B2B social media strategy firm, explains, working with influencers requires a different approach: "Many B2B laggards think they're doing influencer marketing because they have a public relations firm on retainer. They just call it 'Media Relations' and 'Analyst Relations'—let's pay Gartner [a technology research company] $40,000 to show up in their new Magic Quadrant Report" (Fig. 6.6).

"As a B2B influencer in the talent management space, I strive to remain objective and unbiased. It's how I maintain the trust of my audience and the brands I partner with."

-Dan Schawbel (@DanSchawbel)
Partner & Research Director
FUTURE WORKPLACE

DigitalInfluenceBook.com   SHARE THIS

Fig. 6.6   Dan Schawbel

This view is consistent with other industry insiders. A Fortune 500 marketing executive at an American technology firm, who requested their comments remain anonymous, expands: "There's definitely a perception in the industry that analysts can be bought. While we pay influencers for certain activities, it's much more transparent than the old way of doing things." The executive went on to make another key point of differentiation between analysts and influencers: "When it comes to working with analysts, you don't let them see what's behind the curtain, but with influencers you can have a genuine conversation about 'how can we make this better?' and get their sense on the market. You can be a little more open with them."

In Hostelley's view, social has blown up the traditional way of working with external third parties, and Category Influencers have become even more important for B2B firms than traditional media and industry analysts: "Social has allowed…authors, thought leaders, industry executives, bloggers, domain experts all to become influencers. B2B brands are all selling highly specific technology or services, versus broad consumer goods, so taking the time to identify the right influencer is critical to reach the specific audiences who really care about the solutions these firms offer."

He also points out some inefficiencies in how many B2B influencer teams are structured: "B2B social staff historically came from the B2C side where they had experience managing communities, but they don't know how to reach, engage and turn influencers into advocates. Meanwhile, the PR teams know how to work with influencers as people to pitch stories to, but the PR teams are dropping the ball on influencer marketing, because it does not work the same way (media list, spamming, seeing what sticks)—it frustrates journalists and it REALLY aggravates influencers since this is all new to them."

Follower Count ≠ B2B Influence

Chapter 7 introduces a range of factors brands should consider when selecting the right influencers to work with that apply to B2B and B2C. A key takeaway for both sides is number of followers does not equate to scope of influence. However, for B2B its particularly important to consider who exactly an influencer "influences," because if the chief information officers from the top 10 technology firms all follow one influencer, then it does not matter who the influencer's other 10,000 followers are—they are probably a good influencer to collaborate with for a technology firm that wants to sell to CIOs.

Tim Crawford at AVOA makes another key point for the B2B space: "Most marketing organizations focus on amplification, especially when they are first getting started working with influencers. Relationships take time and are harder to quantify, BUT you don't know all the networks I have access to that you want to influence. Especially in the B2B influencer marketing world, there's what's visible via social tools, and then there's the hidden networks (people I have access to and can influence that you can't understand simply by analyzing my Twitter following)—and that's more than just amplification."

Getting B2B Influencer Marketing Right

From Chapters 5 and 6, it should now be clear what some of the major differences are between B2C and B2B influencer marketing, as well as the current challenges both sides are working through today as they advance their influencer engagement programs. For B2B, best practice is to work with influencers through an ongoing "influencer relations" approach, where a brand builds and maintains a collaborative community of Category Influencers that they can then engage at different points in time based on the company's initiatives.

Since there are very few "full-time influencers" in the B2B space, brands need to be cognizant that their influencers wear multiple hats and respect the influencer's need to maintain objectivity in the marketplace. Part of this involves being willing to work with influencers even if they have previously collaborated with competitors. It also means spending the necessary time to identify mutual points of benefit that attract the influencer to want to collaborate on an unpaid basis (at least before the relationship is established).

Additionally, there are very real differences in how B2B brands should treat influencers compared to how they may be accustomed to working with journalists or industry analysts. It may be more difficult to identify the right Category Influencers who speak to the exact customer segment related to a brand's solution set, but the right influencer relationships can have outsized returns when compared to managing relationships with traditional external third parties like media and industry analysts.

Lastly, especially in the B2B space, marketers need to recognize the role of "hidden influence"—the type of influence that cannot be measured by total number of followers. Since there is a relatively limited number of corporate buyers for most B2B solutions, understanding who has the right networks to get in front of decision makers if key to determine whether or not a deal lands their way after a 12- to 24-month sales cycle.

Of course, Chapters 5 and 6 are only a high-level overview of some of the things marketers need to be aware of when it comes to B2C and B2B influencer marketing. What follows is a step-by-step approach to understand how to: identify the right influencers, develop an effective influencer outreach strategy, collaborate with influencers, and measure the return on investment (ROI) (not to mention potential brand risks to watch out for). The next chapter goes deeper, to explain why follower numbers should be taken with a grain of salt and what marketers should really keep in mind when seeking out the right influencers for their brand.

Marketer's Cheat Sheet

- **B2B is as Important as B2C:** While there is a common misperception that influencer marketing is only relevant for B2C firms, influencer relations is critical for B2B brand success as they manage their corporate reputation and stay top-of-mind among potential buyers over drawn-out sales cycles.
- **Unbranded Influencers:** For B2B influencers, their objectivity in the marketplace is the basis for everything they do. They are unable to outwardly favor one brand over another, and they have to take a much more indirect approach with regard to everything they do promote.

- **Dancing Around Dollars**: While B2B influencers cannot directly accept money to promote a brand, the thought in the back of many influencers' minds is always "how will this initial engagement lead to a future paid opportunity?" Brands need to be aware of the dilemma B2B influencers face and work with them to develop mutually beneficial compromises.
- **New vs. Traditional Third Parties**: With most B2B brands being accustomed to working with journalists and industry analysts, they need to be ready to take a different approach with Influencers. Everything from initial outreach to incentivizing influencers to collaborate requires a more thoughtful approach. Influencers expect to be treated as a real person and will not respond positively to spam.
- **Hidden Decision-Maker Relationships**: Given the specific solutions that many B2B brands offer, there is oftentimes a fairly small universe of prospective buyers. Therefore, it is important to consider which influencer has the largest number of relevant followers on social media, but also who might that influencer have in their Rolodex that that brand wants to get in front of.

Notes

1. Huawei. *Huawei CEO & Founder Gives First Ever Interview on Global Corporate Outlook*, 12 May 2013. Print.
2. Disclaimer, Author participated in 2017 Huawei Mobile Broad Forum for book research as a guest of Huawei in London.
3. Israel, Shel. "Why Huawei Is Among the World's Smartest Social Marketers." *Transformation Group*, 2 Dec. 2017, https://transformationgroup.io/huawei-among-worlds-smartest-social-marketers/.
4. Schaefer, Mark. *The Rise of Influencer Marketing in B2B Technology*, 2017.

7

Discover Influencers: Finding the Perfect Match

When Volvo Cars released a new eco-friendly car wash solution in the USA, rather than simply promote the product directly, it marketed the solution around the cause of water conservation. Through its #DrivingDirty campaign, the company challenged consumers to stop washing their car:

> At Volvo, we've always stood for what matters to people. Saving water matters to us all. So whether you drive a Volvo or not, please help conserve water. Turn your dirty car into a badge of honor by writing #DRIVINGDIRTY on it nice and big. Then post it online to get your friends and family involved too.
>
> We also know sometimes you just have to wash your car – like for dates and job interviews. So we introduced Consciously Clean: a waterless carwash solution that makes it easy to keep your car clean and your conscious cleaner.[1]

Its Consciously Clean car wash alternative became the natural choice for consumers wanting to clean their car without having a negative impact on the environment. As a complement to its consumer-focused social media campaign, Volvo also targeted Category Influencers to help amplify its message and reach a broader audience. One of these influencers was Chriselle Lim, a Korean–American mother and fashion influencer based in Los Angeles.

For more than five years, Lim built a loyal following of young, fashionable mothers through consistent production of YouTube videos, Instagram posts, and Facebook updates. Her audience felt an authentic connection to Lim through the thoughtful advice on fashion and raising a family that she

© The Author(s) 2018
J. Backaler, *Digital Influence*,
https://doi.org/10.1007/978-3-319-78396-3_7

shared across social media. This connection catapulted Lim to become one of the top influencers in her category, with more than 1 million Instagram followers and 60 million views of her YouTube videos at the time of writing this book.

However, when Volvo chose to collaborate with Lim, they didn't do enough upfront research to realize that her audience goes to her for fashion and motherhood advice, not tips on how to live a more environmentally friendly lifestyle. So, the company was not prepared for what happened after Lim posted her first promotional Instagram post (Fig. 7.1).

Within moments of posting the promotional content to Instagram, angry comments calling out her lack of authenticity started to come in. Lim rushed to make things right with her followers but the negative remarks kept coming in…

in.pursuit.of.shoes This isn't about being Eco friendly. This is about getting paid by Volvo to shill their cars (and considering the Cayenne you drive…so hypocritical).

bobella I know for a fact the clothing you wear is way more hard on the environment. So it's not REALLY something that's important to you.

mtabajonda2 …It does not make me happy how she's trying to play eco friendly when her outfits and lifestyle are not. I'm going to unfollow.

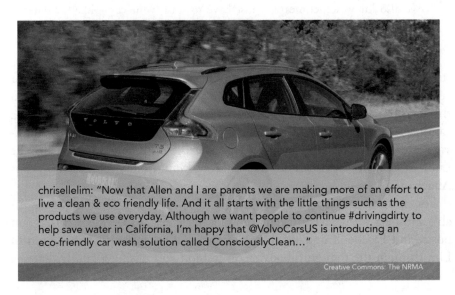

chrisellelim: "Now that Allen and I are parents we are making more of an effort to live a clean & eco friendly life. And it all starts with the little things such as the products we use everyday. Although we want people to continue #drivingdirty to help save water in California, I'm happy that @VolvoCarsUS is introducing an eco-friendly car wash solution called ConsciouslyClean…"

Creative Commons: The NRMA

Fig. 7.1 Volvo Instagram post

tamikaeastley You are becoming such a fake @chrisellelim. This is a sell-out. You don't lead an eco friendly life, that's evident just by looking at your Instagram and the things you do. Please take a look at what image you're creating, because it's clear you're willing to create whatever image you need to make money.

Lim did her best to fight back, and defend her integrity, responding to many of the comments and adding an update to the original post:

After seeing all the comments I wanted to clarify... I'm not saying that I'm perfect and live a complete 100% eco-friendly life, but since I've become a mom I have become more aware of the toxins that we use everyday. I'm only human and striving to become better everyday, and that is the message I wanted to send to you guys. By all means I did not mean that I was perfect...I still have a long ways to go. Thank you for your support and for understanding.

It was too late, though. Lim had broken her audience's trust and Volvo realized they had made a bad investment.

In collaborating with Lim, Volvo neglected to consider the ABCC's (Authenticity-Brand Fit-Community-Content) from Chapter 3—especially the need for Lim to maintain an authentic connection with her community. Her community was quick to draw Lim's attention to a recent post featuring her wasting water while taking a rose-filled bath, and many other photographs of her wearing clothing that had been produced in a non-eco-friendly manner. The Volvo collaboration was not a good brand fit for Lim's audience, and if the Volvo marketing team had invested more in learning their ABCC's upfront, the whole incident could have been avoided.

Getting Started: Target the Right Audience First

For purposes of this book, we'll assume that before a company even considers the "influencer discovery" stage, they have already set specific goals for their company's influencer marketing program.

Goals could include: expanding brand awareness, reaching new target audiences, improving sales conversion, managing their corporate reputation, and so on. The goals will be unique to the company, but there has to be an overarching goal in place that the program is built to achieve. In order to build a resilient, long-term-oriented influencer marketing strategy, a consensus must first be reached among relevant stakeholders about what the company is aiming to get out of its investment in influencer marketing. The last thing a brand should do is "try out" influencer marketing by doing cursory research, choosing someone at random with a large online following, and hiring them to post one-time promotional content about the company. This approach will surely fail (and does happen way more often than brands are willing to admit) (Fig. 7.2).

Gil Eyal, CEO of HYPR, an influencer marketing software firm, tells the story of a luxury watchmaker that did just that. The company built a relationship with an American model who seemed to be a good fit at face value; however, upon closer examination of the model's audience, it became apparent that 70% of her audience is based in South America. When enough attention isn't placed on audience-fit upfront, Eyal says, "Those campaigns fail, and the answer within the company is 'I guess influencer marketing isn't effective for us.'"

Then, there's the time American cosmetics brand Bobbi Brown collaborated with swimsuit model Kate Upton. It makes sense that a well-known swimsuit model would be hired to promote products used by women; however, a deeper dive into Upton's audience demographics reveals 78% of her followers are actually men between that ages of 33 and 39. This doesn't make Upton a bad influencer; it just means that Bobbi Brown isn't an ideal fit for the audience she has influence over. A much better fit was

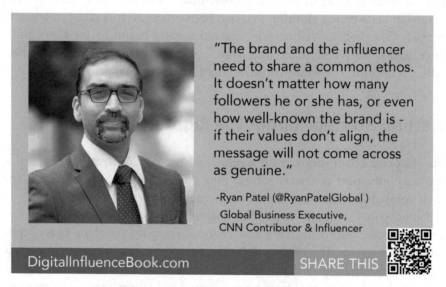

"The brand and the influencer need to share a common ethos. It doesn't matter how many followers he or she has, or even how well-known the brand is - if their values don't align, the message will not come across as genuine."

-Ryan Patel (@RyanPatelGlobal)
Global Business Executive,
CNN Contributor & Influencer

DigitalInfluenceBook.com   SHARE THIS

Fig. 7.2   Ryan Patel

Upton's collaboration with an online game called "Game of War," whose predominantly male audience was much more aligned with her followers' interests.[2]

With the right goals in place, a company needs to think about who exactly they want to reach through influencer marketing. It's about identifying the individuals who have validated influence and can change outcomes with whomever their target audience is—it should start with the target audience and then build back from there. If Volvo really wanted to target eco-friendly consumers with their new car wash solution, they should have worked with a different influencer whose audience was a better fit. With a bit more upfront research, they would have found that Lim was not the right partner because her audience doesn't see her as eco-friendly.

Back in 2011, the French cognac brand Courvoisier launched a marketing campaign to increase its public recognition as an innovative company. Delphine Reynaud, then head of Courvoisier's innovators community at White Label UK, the brand's creative agency, advised against offering the influencers gifts such as Prince concert tickets because as she explained to the team: "If you want to foster a dialogue about our brand as being innovative, how does the experience of a Prince concert get them talking about our brand in that way? Yes, it's a 'cool' experience, but it's not consistent with our goals for the community."

"I still think that there's a lot of bias – that's always the challenge of not having confirmation bias and going into get validation for what you wanted to see when you started an analysis in the first place," says Chris Gee of the global public relations firm Finsbury. He says marketers often get into trouble when they are too focused on the tactic of influencer marketing and the tactical outcome, as opposed to truly going into it with the mindset of discovery.

Marketers need to go in with the mindset, "What key insights can I understand about my customers?" Gee says. "The challenge is a lot of marketers will come into an influencer audience analysis, look at the data and say, 'How can I confirm that we want to use Scarlett Johansson as an influencer?' They'll completely miss shifting attitudes about their product, potential new products, [and] product innovation, because they're so hypermotivated to find out how they can make this deal with Scarlett because they had a great meeting with her rep, and know they can make it happen." While marketers have access to much more data than before, they don't always have the discipline to understand how to analyze the data in an objective manner.

Next: Target the Right Influencers Who Influence That Audience

75%

"Seventy-five percent of marketers believe finding the right influencers is the most challenging part of establishing an influencer marketing program," according to eMarketer.com

After defining target audience, marketers can then determine who is influential over that audience. But it's not always easy to identify the right influencer. Seventy-five percent of respondents participating in a 2015 eMarketer.com survey ranked influencer discovery as the top challenge they face in establishing an influencer marketing program.[3] In the years since this survey, a series of tools and resources have become available that make it easier for marketers to identify the right influencers, which we'll explore later in the chapter, but it still takes much more than a simple database search for a company to identify the best influencers.

Identifying the right influencers that align to a company's target audience can be time-consuming and complex, but the upfront research is a must. "At the end of the day, brands need to be more strategic and ask influencers for media kits [and] Google analytics to ensure that the demographics align with the products or brands, but the most important thing is the research—no brand, campaign, engagement should ever start without researching to ensure alignment between what's on brand for the influencer and what's on brand for the company," argues Ruben Ochoa, an influencer marketing specialist based in Los Angeles.

Remember from Chapter 3, influencer marketing practitioners generally look at the following three factors to measure an influencer's potential impact: Reach, Relevance and Resonance:

- *Reach*: The total size of an influencer's audience across all social platforms measured by followers, subscribers, traffic, etc. Think of reach as what defines an individual as a Celebrity Influencer, Category Influencer, or Micro-Influencer as related to the earlier section of this chapter.
- *Resonance*: Engagement between the influencer's audience and the content they produce measured by shares, likes, views, comments, retweets, etc. Resonance is important, because influencers need to be able to demonstrate

that their community is more than just a number, it's engaged and interested in their content.

- *Relevance*: Content-topic match ensuring that the content produced by the influencer is aligned with a consistent set of topics that is of interest to the influencer's community. From a brand's perspective, relevance also relates to how closely an influencer's community matches up to the brand's target audience, as well as how closely the influencer's content aligns with key topics that the brand wants to be associated with.[4]

Influencers do not necessarily need to score high across each of the "Three R's." Companies may intentionally seek out influencers with high scores for Resonance and Relevance, but with limited reach, depending on their goals. "For instance, a lot of the industry analysts I have invested in typically don't have more than 150 followers, but when you dig into the analyst's audience they're all very senior executives at Fortune 500 companies," explains Chris Purcell, manager, influencer marketing for Hewlett Packard Enterprise. "It's a mixture. You really need to do your homework to understand just because someone has a smaller number of followers, if their audience is engaged and is your target audience, then it still may be worthwhile to invest in that influencer relationship."

Beyond the "Three R's," influencer marketing professionals assess a range of other factors as well. Konstanze Alex-Brown, head of Global Digital Communications at Dell, explains his team's approach: "As a starting point, we use algorithms offered by a variety of companies and sometimes work with an outside agency to find these micro-influencers," but that's only the first step. "But if you stop there, you can pretty much write off your program, because it's not going to work … We start with a long list of influencers, but then I tap the intelligence of the team and ask 'Hey, do you know this person? What do you think'? That's how an influencer gets to the top."[5]

On the agency side, Simon Kemp, chief executive at Kepios, introduces his approach: "We have a framework for identifying influencers in a particular field or category. It depends on what the client is trying to achieve, but a certain amount will look at the reach, the credibility, the frequency, the strength of opinion backed up by rationale." His framework also includes a scoring system that accounts for both the audience quality and the content quality. "It looks at a variety of different variables and then comes to a score that allows us to compare different individuals even if they're doing very different kinds of things. For example, how would you quantify the value of a very passionate advocate within a relatively small niche, versus a very prolific influencer with great reach, but not a massive expertise in the particular topic that we're talking about?" (Fig. 7.3).

"Marketers should consider five factors: engagement, face recognition (fame), quality of content, competitor campaigns and price."

-Senior Director of Influencer Strategy & Talent Partnerships HEARST MAGAZINES DIGITAL MEDIA

DigitalInfluenceBook.com   SHARE THIS

Fig. 7.3 Brittany Hennessey

Influencer discovery gets more complicated the bigger the brand and the more global markets the brand operates in. Coty Inc. is a $9 billion American beauty company that operates in countries all around the world. Bruna Scognamiglio, vice president global influencer marketing for Coty's Gucci Beauty division, describes the two ways her team identifies the right influencers: "The first is project-oriented. For example, when launching a product where we really wanted to elevate its positioning as a high-end item, the brief for influencers was to go very niche, even if the product was eventually a 'beauty commodity'." Each influencer who participated in the project needs to influence a certain kind of people, so the influencer helps tailor the message for unique audiences even though the product itself has traditionally been viewed as a mass-market product.

For their second approach, depending on the project, they try to identify in which of the countries there is the relevant business need. For example, Scognamiglio says, "I might need to go find an influencer who is stronger in the US versus someone whose audience is more balanced globally or maybe some focus on the Middle East – you need to understand where you want to have an impact."

For Scognamiglio's product category, the influencers she works with need to fit with the specific brand, because every fashion brand has its own positioning in terms of how its consumers perceive it and its relevance: "What is relevant for Converse lovers isn't necessary what is relevant for a luxury lover as well what is relevant for Gucci beauty lover for instance is not necessary relevant for Chanel ones. Nowadays each brand has its own equity and point of difference and tells a different story."

So "brand fit" is very important to the selection of influencers. Influencers need to be relevant and authentic, as well as influential for a company's target consumer. Scognamiglio has systems to monitor influencers: "We track who…the most up-and-coming influencers [are] – we track whose audience is growing country by country, we try to do a qualitative analysis to understand if there is a switch in the way they behave (in the last two to three years it has been very stable) and then we try to understand who are the ones in specific categories are the strongest."

This ongoing monitoring requires a lot of effort. "We try to build relationships based on the numbers we get every six months, but then it's the project itself that determines who we're going to work with, because you need to take your creative brief, understand what you want to do and then activate the right influencers." In working with influencers, marketers need to map out the whole process, starting with influencer discovery, and determine how they will manage it. There are situations where in-house management makes sense, but then there are also agencies and various software providers that handle different aspects of influencer marketing management.

3 Paths for Influencer Management: In-House, Agencies and Software Solutions

The rapid development in recent years of technology-enabled influencer marketing means very few companies have mature, established teams in place to manage it. Brigitte Majewski, vice president of business to consumer marketing research at Forrester Research, said it best: "The current influencer marketing landscape across brands, agencies and software vendors is a mess."

Brands rush to hire junior-level self-proclaimed "influencer marketing specialists," or they simply add "influencer marketing" to an existing employee's long list of other responsibilities. Meanwhile, various agencies have popped up that specialize in all aspects of influencer marketing, often managing their own dedicated roster of influencer talent. Pitching both agencies and brands are representatives from literally hundreds of influencer marketing software solutions, ranging from talent marketplaces to influencer identification tools to influencer CRMs (Customer Relationship Management software repositioned to manage influencer relationships). No matter whether a brand manages their influencer marketing efforts internally or with external support, there are a series of key points that marketers need to keep in mind when defining their influencer identification approach.

Path 1: Manage Influencers In-House

Within brands, influencer marketing tends to sit in varying departments from company to company. Sometimes, this variation is intentional, based on the goals of the brand's influencer program—for example, if the goals are focused on corporate reputation, then management might sit in corporate affairs, but if the goals are more aligned to driving demand, then the marketing team may be a better fit. Oftentimes, influencer marketing management decisions tend to be made arbitrarily.

Influencer marketing management differs so much across companies because there is a general lack of established best practices in place, so brands are forced to experiment and figure out what works best for them. They do so with limited resources, because "for a chief marketing officer, any decision to invest in one area comes at the expense of investing in another area," explains Kimberly Whitler, Assistant Professor at the Darden School of Business and former CMO for several consumer brands. This means as influencer marketing is the "hot new thing" for many companies, it does not receive the necessary attention it deserves, because existing human resources, strategy, and budgets are tied to more established and well-proven business activities.

When IBM Watson Customer Engagement began developing its influencer relations capabilities, its social media team spent a lot of time on influencer identification, including working with a few agencies to help them with that, according to Amber Armstrong, who led IBM Watson Customer Engagement influencers at the time. "As we built up the necessary skills internally, we've been able to hire and maintain people on staff who do a lot of this identification work for us." By successfully identifying the right influencers with the help of an outside agency, the in-house team was able to later make the case for additional resources to bring on more specialized staff to manage the process on their own.

The reason why brands often rely on agencies or software solutions to help identify influencers, especially when starting out, is the process of influencer identification can be extremely time-consuming and inefficient or as one interviewee how manages influencer marketing for a large B2C firm describes: "I spend hours every day searching for new influencers online, reviewing their content, reading follower comments, checking for mentions of our competitors, digging into vanity metrics like follower count…it should be a full-time job, but it's one of several competing priorities I need to balance" (Fig. 7.4).

"Brands should consider working with agencies, especially on an integrated influencer marketing strategy that ties content development together with influencer marketing program implementation."

-Lee Odden (@leeodden)
CEO | TOPRANK MARKETING

DigitalInfluenceBook.com   SHARE THIS

Fig. 7.4 Lee Odden

Path 2: Hire an External Agency

In many cases, agencies can help brands identify influencers by leveraging the agency's existing influencer relationships, a more efficient method for a brand than taking on all of the efforts in-house. Agencies come in all shapes and sizes, ranging from large, global agencies to niche industry-focused, to country-specific specialists—there is likely an agency partner out there to satisfy the needs of just about any brand. Influencer marketing agencies generally have exclusive talent or pre-existing relationships with an extensive network of tested influencers. This relationship-based approach means agencies often have more information about influencers than what can be discovered through online research alone.

Beca Alexander, founder and president of the influencer marketing agency Socialyte, regularly fields questions from brands to identify influencers in her network that are approaching important life milestones.

Some examples include:

Who in your network is going to get married in the next six months?
Who in your network is planning to buy a house?
Who in your network is about to have a baby?

"Those are insights that online research or technology platforms will never be able to track. That requires the established influencer relationships that we have and cultivate over time," she says.

Brands should choose agencies carefully to ensure they are really bringing something special to the table as a result of their experience and influencer relationships. Agencies can easily subscribe to one of the influencer marketing software solutions introduced in the next section and then connect brands with influencers who they have no prior relationship with. This could still be advantageous for smaller brands that may not be able to afford costly software subscriptions, but larger brands may be better off directly purchasing the software themselves. Other critics of influencer agencies contend: "The big problem is, they don't operate much like a traditional talent management company. They don't provide insurance in case their talent doesn't deliver or anything,"[6] according to an anonymous executive interview published by Digiday.

Larger brands also need to make sure they are working with agencies for the right reasons. Fashion influencer Heidi Nazarudin describes a recent meeting she had with a brand: "This one company I work with has a separate agency that manages its PR efforts which is normal. There are so many people in the room when we meet, which is fine, but this particular agency clearly had no clue on how to assess and manage influencers – it's like the blind leading the blind in the best-case scenario." An outside agency can be beneficial provided the agency has a track record of delivering campaigns that have concrete deliverables and actionable metrics. However, brands need to make sure that the internal decision maker working with these agencies is in-tune and aware of what a great digital campaign is all about. This decision maker, usually the digital manager or social media director, needs to be able to grasp what makes a campaign ideal for a brand's marketing goals, and what may not be relevant.

Path 3: Purchase Influencer Marketing Software

Both brands and agencies may choose to subscribe to influencer marketing software solutions to help with influencer identification and other aspects of influencer marketing management. In recent years, hundreds of software solutions have popped up, offering different types of functionality. To simplify the universe of software solutions, they generally fit into the three buckets of influencer discovery, influencer marketplaces, and influencer relationship management.

Influencer Discovery

- *Purpose*: An algorithmic database allowing users to automatically discover the right influencers based on robust filtering criteria matched to influencer audience data
- *Examples*: HYPR Brands, Affinio, Onalytica

Influencer Marketplaces

- *Purpose*: A marketplace to streamline the process of brands connecting to influencers
- *Examples*: Tribe Group, IZEA, FameBit

Influencer Relationship Management

- *Purpose*: Most similar to a sales CRM, it tracks and monitors engagement with influencers
- *Examples*: Traackr, IFDB.com, GroupHigh

Out of the three types of software solutions, Gil Eyal of HYPR feels marketplace platforms are in the most trouble: "They've spent an enormous amount of time and energy to build a roster that they can provide access to," but the challenge for marketplaces is their roster participants are often signed to many other marketplaces. "While a brand or agency can use these marketplace platforms to reach out to influencers directly, the platforms themselves do not really offer anything unique."

Eyal feels marketplaces are trying to get influencers the most money possible so that the influencers stay with them. The marketplace in turn gets a bigger cut of the deal as well. He argues that brands are in a position where these platforms just make it more expensive to do something they could otherwise accomplish by reaching out to the influencer directly as an established brand.

He is not alone in his opinion of influencer marketplaces. "The problem is, the very first conversation an influencer and brand have on these platforms is 'How much are you going to pay me?' But how do you build an authentic relationship based purely on a dollar figure?" Christian Damsen of Traackr argues marketplace companies are leading to poor habits on the brand and influencer sides, respectively. "These transactional-based relationships will not be sustainable or scalable over the long run."

Justin Szlasa, chief executive officer of IFDB.com, an influencer relationship management software, feels that many of the available software solutions for influencer marketing are priced out of reach for smaller brands: "This is because these software firms tend to be VC-backed and face a lot of pressure from their investors to grow fast by signing big-time deals with large, enterprise clients." According to Szlasa, companies of all sizes need access to the right tools to track and engage members of their influencer community and, more importantly, to maintain alignment between different internal stakeholders and not over-communicate or send conflicting messages to the brand's influencers.

Despite the many options available on the market, there is no market-leading influencer marketing software brand, because the environment continues to evolve. "Influencer marketing has developed primarily through well-known social channels – we're starting to see the patterns so we can now apply some sort of software to map to those patterns at scale," explains Scott Brinker, a marketing technology expert. "However, the challenge is the environment that they're trying to establish these patterns in just continues to shift, which makes it really hard for the largest software players to aggregate the smaller niche players into their core products and keep up with the pace of change."

For now, this means that influencer marketing software is best suited for brands and agencies that require a technology solution to identify influencers and manage existing influencer relationships. Brands need to be cautious of any software that claims to offer an end-to-end influencer marketing suite, because the current environment continues to evolve. Even within the existing set of available software for influencer identification, no single provider is perfect. Brands and agencies should use search results as a jumping off point for deeper influencer research and not take the results at face value. Ultimately, influencer marketing requires brand representatives (either in-house or agency) and influencers to interact, which is highly variable, and at least for now not scalable for a software solution to come in offering a fully automated solution.

Discover Influencers: What It Really Takes to Identify the Right Influencers

Before influencer discovery can begin, brands need to establish clear goals for their influencer marketing programs. These goals will vary depending on the organization, but goals can include areas such as expanding brand awareness, reaching new target audiences, improving sales conversion, and managing corporate reputation. Developing these goals requires marketers to think long and hard about the specific audiences they intend to impact through collaboration with influencers. With the foundation of a sound strategy and target audience in place, marketers can then begin to identify the right influencers for their program.

Brands can take three paths to identify influencers: They can manage the entire process in-house, hire an agency, or purchase influencer marketing software. More often than not, a brand will likely adopt some combination of the three. For example, when just getting started, a brand may hire an agency to help develop capabilities for influencer identification prior to managing the program in-house; or both brands and agencies may choose to also purchase influencer marketing software to help streamline the influencer identification process.

Of course, identifying an influencer is only the first step in a series of actions required to build a connection between a brand with that influencer. Engaging influencers to cultivate authentic relationships is the next step and the focus of Chapter 8. Remember, just because an influencer looks like the "perfect fit" for a brand, this doesn't mean that he or she will want to collaborate. Next up is a series of "make or break" steps in the influencer relationship building process that brands will not want to miss.

Marketer's Cheat Sheet

- **Target Audience First**: Many brands like to start with a "cool influencer" and then find a way to reverse engineer analysis to justify why he or she is a good fit. This approach fails every time—companies should begin influencer discovery by first defining a clear set of goals and a target audience.
- **Look Beyond Vanity Metrics**: A large number of followers do not automatically make an influencer the right choice. Remember the influencer ABCC's, and spend the time to research other qualitative factors including the quality of engagement, face recognition (fame), quality of content, industry fit, and so on.
- **Choose the Right Path**: Depending on brands goals for their influencer marketing program, they can choose to manage everything in-house, hire an agency, or purchase influencer marketing software.
- **In-House Management**: This path gives marketers the most control, but they may not have the necessary resources and budget to manage the program effectively—especially when first getting started. Consider working with an agency to receive training and leverage their network of influencers or expensive software subscriptions.
- **Influencer Agencies**: This path gives marketers less control; however, if the agency is a good fit, brands can get even better insights than what can be found online. For example, brands can learn about specific influencers in the

agency network who are going through important life events that relate to the brand. However, agency quality varies and brands need to ensure they are getting what they pay for, and not paying way too much for someone to blame in the future if the initiative fails.

- **Influencer Marketing Software**: This path is more of an enhancement to the first two. Brands and agencies can purchase influencer marketing software to help identify influencers or manage existing relationships. While there are seemingly endless software solutions available, none are perfect and the environment changes every day.

Notes

1. Volvo Cars USA. "Driving Dirty." *Driving Dirty | Volvo Car USA*, www.volvocars.com/us/shopping-tools/additional-choices/events/driving-dirty.
2. HYPR. "The Power of an Individual—Part 2—HYPR." *HYPR Influencer Marketing Platform*, 2017, hyprbrands.com/blog/the-power-of-an-individual-part-2/.
3. "Marketers Pair Up with Influencers—And It Works." *EMarketer*, 9 July 2015, www.emarketer.com/Article/Marketers-Pair-Up-with-Influencersand-Works/1012709?ecid = MX1086.
4. Solis, Brian. *Influence 2.0: The Future of Influencer Marketing*, 2017.
5. Schaefer, Mark. *The Rise of Influencer Marketing in B2B Technology*, 2017.
6. Pathak, Shareen. "Confessions of a Social Media Exec on Influencer Marketing: 'We Threw Too Much Money at Them'." *Digiday*, 12 May 2016, digiday.com/marketing/confessions-social-media-exec-no-idea-pay-influencers/.

8

Engage Influencers: Developing an Effective Outreach Strategy

It's a typical day for C.C. Chapman, an American content marketing Category Influencer. As he scrolls through his Twitter feed as usual, a notification from the pasta sauce brand Ragu catches his attention. Ragu traditionally targets mothers in their marketing efforts, but, as part of a new campaign, the brand wants to involve dads and help them become more active in the kitchen. To do so, the brand identifies a series of dad Category Influencers and uses Twitter to reach out to them. The challenge, at least in Chapman's case, is he found Ragu's outreach to be really offensive.

This is what Chapman saw when he checked out Ragu's Twitter message:

@cc_chapman do your kids like it when you make dinner?

His immediate reaction was shock. "I've never interacted with Ragu on Twitter before," he said, "and honestly don't buy their product, so when I got the message from them linking to a video, I wondered what it was. A quick look at their Twitter stream showed me that they had spammed a bunch of dads with a link to the same video." This is what Chapman saw when he reviewed Ragu's Twitter stream:

@playgrounddad who makes dinner in your house? Mom or dad?
@marcdecaria how does dinner by dad stack up?
@dwaynereaves do your kids like it when you make dinner?

In response to Ragu's spam outreach, Chapman wrote a series of blog posts for his audience explaining his perspective, starting with one entitled

© The Author(s) 2018
J. Backaler, *Digital Influence*,
https://doi.org/10.1007/978-3-319-78396-3_8

"RAGU HATES DADS.[1]" He writes about why he takes particular offense to the blanket outreach:

> As the person in my household who does **all** of the shopping and **all** of the cooking, I took offense to this video. Implying that dads can only cook the simple things and Ragu is somehow going to help make that easier – give me a break!
>
> I'm sure there are plenty of couples out there [for whom] this might be true, but once again we have a brand who has decided to only focus on the mom side of the parenting equation and play into the stupid stereotypes that dads get pegged with all the time.

In his final post on the topic, "MY FINAL WORD ON RAGU,[2]" Chapman shares takeaways from his phone call with a brand manager at Unilever, Ragu's parent company at that time. On the call, Chapman shared a long-standing business and messaging strategy: *If you want any other audience beyond your target market to be interested in your product, you must make them feel welcomed.* He explains, "When my first interaction with a brand is an @ spam on Twitter…and when I engage and yet see nothing to welcome me…that is a turn-off. Follow that up with a video that insults me and my friends? Yeah, not exactly the welcoming committee I would have expected."

Ragu's negative experience engaging Chapman highlights the importance of tailored influencer outreach. It is unlikely that anybody within Ragu or their agency reviewed Chapman's blog content, as they would have realized that he is an active participant in his household who could take offense to outreach that labels him as a "dad who doesn't contribute." Further, because Twitter @'s are all publicly available on Ragu's Twitter feed, it was very easy for Chapman to find out that the brand was spamming multiple dad influencers with similar messages. By not taking the time to cultivate a relationship with each influencer individually, Ragu not only did not get the campaign results they were looking for, but they ended up getting a lot of unwanted, negative attention for their influencer engagement blunder.

It is easy for marketers to get excited after they identify the perfect set of influencers using the best practices introduced in this chapter. It can take a long time to find influencers with the exact right audience and a content focus that is aligned to the brand's goals, so it is only natural for marketers to get excited and want to reach out right away to get the conversation started. However, patience is key—just because an influencer is a good fit for a particular brand, it does not mean the influencer will automatically want to collaborate with that brand. Therefore, marketers need to take a step back and assess the best outreach strategy to build an initial relationship with their target influencer.

Things to consider when developing an influencer outreach strategy include:

> Does the brand intend to pay influencers?
> Will the brand–influencer relationship be one-off or ongoing?
> Who will contact the influencer on behalf of the brand?
> What message and channel is most appropriate for the initial outreach?

Many brands underestimate the challenges they may face when reaching out to an influencer for the first time. Oftentimes, the mentality is: "We're famous brand X, of course they will want to work with us." But as this chapter reveals, it is all about the details—it is not about how famous a brand is, it is about how the influencer's first interaction with the brand on an individual level is received. Effective outreach is everything.

Incentivizing Influencers: To Pay or Not to Pay?

One aspect where influencer outreach varies is based on how a brand intends to incentivize the influencers it wants to engage. Depending on whether a brand plans to pay influencers, offer non-monetary incentives or adopt some combination of the two, the way they build the initial relationship with influencers differs—influencer marketing requires different muscles depending on the circumstance.

Take the following two situations: Brand X wants to work with a YouTuber with 10 million subscribers to launch a new product, and Brand Y wants to work with a policymaker to influence industry legislation. The outreach approach in each case will vary significantly—for a YouTuber, outreach messaging will likely involve details about what Brand X is willing to pay; brands cannot pay public officials, so the basis of outreach communication for Brand Y needs to be around identifying mutual points of interest between the brand's goals and the policymaker's top priorities.

Organic Outreach: Identify Mutual Points of Interest

Brands may choose not to pay influencers for a variety of reasons. In the policymaker example above, there are simply situations when brands are not allowed to exchange funds. The same goes for engaging influencers for charitable causes, as well as many influencers in the B2B space. Since B2B influencers are often employees of established companies, or run their own businesses, they are not able to accept payment for promoting other brands. Then, there are cases when brands have made a business decision to not pay influencers, even though they are legally and ethically able to do so. Whatever the circumstance that prevents a brand from paying an influencer to collaborate, it is critical for brands to identify mutual points of interest prior to reaching out to an influencer, because influencers are regularly bombarded with requests—no influencer is sitting around eagerly waiting to help any brand that happens to reach out to them (Fig. 8.1).

How can brands provide value to influencers without paying cash? It may not always be straightforward, especially before any relationships exist between an influencer and a brand. Think back to Chapter 3—influencers have a series of motivations for doing what they do that all tie back to the influencer Authenticity-Brand Fit-Community-Content (ABCC). At the most basic level, influencers want to provide increasing value to their community in the form of content and engagement and grow the number of people they have influence over. In framing influencer outreach, brands should think about what they can offer that could help the influencer engage his or her existing audience and help broaden awareness among relevant audiences about the influencer to increase his or her followers.

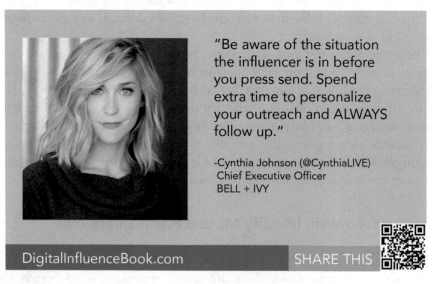

"Be aware of the situation the influencer is in before you press send. Spend extra time to personalize your outreach and ALWAYS follow up."

-Cynthia Johnson (@CynthiaLIVE)
Chief Executive Officer
BELL + IVY

DigitalInfluenceBook.com   SHARE THIS

Fig. 8.1   Cynthia Johnson

According to Shonali Burke, a growth strategist, influencer marketing expert and communications professor at Johns Hopkins University, "If you're genuinely trying to help influencers, then it does come back to you. And maintaining and growing these relationships over time is what leads to something extraordinary." Brands can help influencers add value to influencers' communities and increase targeted followers in a wide variety of ways. For starters, brands can offer exclusive access to company executives, new products, unique content, or events. For example, if a Category Influencer writes a blog about a related industry, offer an exclusive CEO interview to discuss new market trends or offer access to a corporate video studio to film professional video content. A higher level of access will help influencers develop new content or engage their community with exclusive access to ideas or products that they cannot get anywhere else. Unique experiences including conference access and international travel opportunities are another way to help influencers.

Huawei, a Chinese technology company, does not pay its influencers, but it does offer exclusive experiences for members of its influencer community. "What I will pay for is influencers to travel around the world and learn more about Huawei at events where our firm is either the major presenter or at one of our conferences," explains Walter Jennings, who leads influencer relations.

He shares the example of the Mobile World Conference in Barcelona, which is one of Huawei's largest events: "This past year I had a delegation of 24 influencers from 12 countries who attended. Our request of them was in exchange for business class travel and five-star hotel accommodations they would live cover what they saw at the event without any editorial guidance from Huawei (I don't ask to review anything) and I asked for one long-form blog post based on an element of what they saw at the event and their impressions of Huawei." Live events and online events are a great way to engage influencers since they present opportunities to develop exclusive content, and get more exposure to new audiences to help grow the influencer's following (Fig. 8.2).

Amber Armstrong of IBM Watson Customer Engagement shares her view on how brands can drive organic influencer engagement: "When companies create transactional relationships instead of actual relationships, then you get what you pay for." Amber recommends giving influencers exposure to executives, inviting them to employee-only events and co-creating content with influencers (e.g., webinars and white papers) to help influencers gain exposure to new audiences. B2B brands like IBM and Huawei are certainly more likely to engage influencers with these types of non-monetary incentives, but particularly for B2C brands, there are many cases where "pay to play" is the only way.

Pam Moore ✓ @PamMktgNut · 15h
No sorry, I can't cone spend 2 days travel + 3 days at your event as a free influencer for "exposure". Value my time & we value your brand.

💬 2   ↩ 3   ♥ 16   ✉

DigitalInfluenceBook.com   SHARE THIS

Fig. 8.2   Pam Moore Tweet

Paid Outreach: Avoid Spending Too Much, and Getting Too Little in Return

Reaching out to influencers to propose paid forms of collaboration can be both more straightforward and more complicated at the same time. For certain consumer-facing industries, including health and wellness, fashion and beauty, and travel and lifestyle, paid influencer collaboration is the norm. This means influencers focused on these industries tend to be more receptive to outreach offering paid opportunities.

However, the greatest challenge for marketers who want to pay influencers is there are no standard benchmarks or best practices for how much to pay. A Category Influencer could quote $300 for a social media post, while another Category Influencer with similar following could quote $3000, and a Celebrity Influencer may request over $100,000 to post any form of branded content. The rates that most influencers quote brands are mainly arbitrary, most often based on what another brand has previously paid them, or what they think their time is worth. Brands need to be extremely careful here, because high influencer fees do not guarantee winning results—they should spend time applying best practices from Chapter 7 to make sure the influencer is a good fit, with the greatest likelihood to yield a positive outcome, prior to considering contacting them.

Ashley Villa, chief executive officer of the influencer talent agency Rare Global, works with many of the world's biggest consumer brands, and even she gets frustrated by outreach requests from some of these companies. She thinks brands need to develop relationships with talent first and foremost, but the problem is most brands will never get the chance to.

Her advice to potential brand collaborators is this: "Brands should spend time developing relationships with content creators. Our content creators only partner with brands who have values and ethics they can stand behind. Authenticity is so important in the influencer space. We want our audience to trust that our content creators are marketing products that are vetted by them. The best way to achieve this is for brands to have an open dialogue with our team. We don't upload posts in exchange for product. It lacks originality and does not have long term value." When influencers are represented by an agency, the "product-for-post" request from brands is even less attractive, because influencers can't pay their management in the form of free product giveaways.

Paid influencer collaborations do not stop with B2C companies, because there are situations when B2B influencers should be paid for their efforts. B2B tends to seek out influencers who they can build long-term organic partnerships with; however, according to Michael Krigsman Industry Analyst and host of CXOTalk, "In the back of their mind, B2B influencers are thinking about 'how am I going to make money from this?'"

There are three ways by which B2B influencers are generally compensated for their efforts:

- Direct cash
- Notoriety from a large company promoting the influencer and building his or her credibility/social following
- Payment for services—the brand hires the influencer for his or her expertise to work on a project relevant to their expertise at a future date

Since most B2B influencers will not accept cash to directly promote a brand (e.g., being paid for attending an event or for promoting a tweet), they tend to accept payment for anything where the influencer invests his or her time, including delivering keynote speeches at an event, leading webinars, and developing custom content specifically for the brand. Krigsman feels that it is a delicate balance on both sides: "I can assure you payment is an area that makes people very uncomfortable. There's a lot of exchange of goodwill. Companies spend a ton of money to treat influencers as VIPs and influencers spend their time. There's a lot of feel good, and it's genuine for sure, but there's more than meets the eye under the surface."

To pay or not to pay is the first lens to assess how a brand will work with an influencer. Brands also need to determine if they plan to work with an influencer for a one-off engagement or if it would be more beneficial to build a long-term ongoing partnership. There are advantages and disadvantages to both, but knowing this ahead of time will help brands with their influencer outreach strategy.

Engagement Duration: Short-Term Fling vs. Long-Term Relationship

Engagement duration is another factor that brands should consider before reaching out to an influencer for the first time.

Short-Term Influencer Engagements

Short-term influencer engagements are generally used for paid influencer collaboration. After the collaboration occurs, the brand and influencer go their separate ways. Simon Kemp, chief executive officer of Kepios, a marketing consultancy based in Singapore, likens one-off, short-term engagements to the movie *Pretty Woman*. Instead of hoping that Influencers will fall in love with a brand because of the financial compensation they receive, it is "way better to fall in love organically if you want it to be a long-term strategy." The reality is, sometimes brands need a short-term boost (such as when they are launching a new product), but the key is in how the brand manages the influencer relationship afterward—it is a waste of resources to let a good influencer relationship disappear once a one-off campaign concludes.

Reb Carlson, associate account director at the digital agency Wunderman, shares her experience: "One of my previous clients in the quick service restaurant (QSR) space (they launch new food products and beverages all of the time) might have a different influencer engagement approach for each of its campaigns depending on seasonality or market trends."

She describes how the QSR brand might work with a group of influencers to launch a new type of healthy iced tea. For that campaign, the brand may start by choosing to work with a group of health-conscious influencers just for this one initiative. Even though it starts as a short-term engagement, by managing the relationship effectively throughout the campaign, "we may identify influencers that are more relevant in terms of their association with brand – their personality matches the brand personality and they can speak to a whole suite of products then we might want to have more of a long-term relationship with them."

Long-Term Influencer Engagements

Then, there are long-term influencer relationships, where the brand works with influencers multiple times over the span of a couple of years. Carlson feels that consumer brands are more familiar with doing short-term engagements, but that "people who really see influencer marketing as part of a long-term strategy for their business are looking to invest in long-term relationships with people, especially if they took the time to find a quality person who has real followers and creates great content are influencers who you want to hold onto relationships with."

Nicole Smith, global brand and innovation communications manager for Intel, explains their long-term-oriented approach: "We made a strategic decision…to focus more on long-term relationships. As a result, we have something we call an Influencer Management Review Committee, which is supported by our CMO. Anybody at Intel who is looking to engage with digital or social influencers, has to run their proposal through this committee. And one of the key questions for their proposal is, how will you maintain this relationship after that campaign? And if they don't have an answer for that, we don't allow that to go through. This is our answer to driving a culture in the company that focuses on relationships, on ensuring that the influencer has a good experience, that we are meeting all FTC guidelines and to enable multi-discipline engagements."[3]

Amanda Duncan who leads customer lifestyle influencer relations at Microsoft has the following advice for brands: "Focus on a long-term approach rooted in a two-way dialogue. It's often the phases between campaigns and events that allow you to have in-depth conversations, get valuable feedback and really gain a deeper understanding around what matters to your influencers. Investing this time and valuable resources builds credibility. This credibility and trust with an influencer is key to ongoing success."[4]

Huawei bases the success of its entire influencer relations program on long-term engagement: "The number one way we track the success of this program is based on the number of interactions with select influencers over the course of a year," Jennings says. Huawei has developed a relationship-tracking formula to track key moments, such as when an influencer visits headquarters for a campus tour and to meet with executives. "Coming to the Shenzhen campus needs to be worth five points, interviewing an executive is worth five points, and the fact that they wrote positive about us on social media has to be worth about two points, etc. During the course of the year, we find the more we engage influencers, the more they understand, the more correct and broad their coverage becomes – we want these people to keep coming back and they have a comfort level in their relationship with the company."

Influencer Outreach: Who, What, and How

Understanding whether influencer collaboration will be paid or unpaid, or short term or long term, will help inform the appropriate outreach strategy (Fig. 8.3). Prior to contacting influencers, organizations should align on who is responsible internally for influencer outreach. There is also a series of actions that an individual should perform prior to communicating directly with the influencer, and then, there is a collection of best practices to follow when crafting the first message to an influencer.

WHO Is Reaching Out?

In most companies, multiple departments aim to work with influencers in different capacities. Marketing may work with influencers to launch a new product, public relations may work with influencers to amplify a new

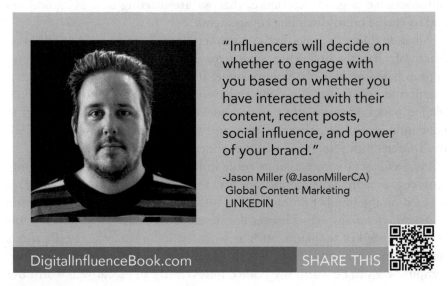

"Influencers will decide on whether to engage with you based on whether you have interacted with their content, recent posts, social influence, and power of your brand."

-Jason Miller (@JasonMillerCA)
Global Content Marketing
LINKEDIN

DigitalInfluenceBook.com   SHARE THIS

Fig. 8.3   Jason Miller

company announcement, and corporate affairs may work with influencers to manage the company's corporate reputation—they are not structuring influencer outreach to attain a common goal. "A lot of clients that we work with sometimes have 10 different people in the same organization who contact the same influencer with slightly different messages," explains Tim Williams, chief executive officer of influencer marketing software firm Onalytica.

Having multiple voices reaching out with competing messages is one of the worst things to do when building a relationship with an influencer. For the influencer, the brand is one voice. It does not matter which person in which department is reaching out, or if an agency is reaching out on behalf of the brand—they all represent a single brand in the eyes of the influencer. Brands should decide on a point of contact to interface with influencers (ensure the person is of equal seniority to the majority of influencers he or she is reaching out to) and prevent anyone else from reaching out to influencers without at least keeping that individual or agency in the loop.

WHAT Happens Prior to Outreach?

Prior to outreach, brands should make an effort to "warm up" the influencer over a sustained period of time in the hopes of building goodwill. The Golden Rule of social media holds true—always give first. Brands should add value to the influencer by sharing their content, promoting their work through brand channels, and commenting on their work. (Organizations should take the time to come up with something thoughtful and personal, however, avoiding generic statements such as "great post" or "loved this.")

Chris Purcell, manager, influencer marketing manager at Hewlett Packard Enterprise (HPE), describes how he engages influencers: "When I find someone I typically do research on them. I either found them or someone found them for me. Because we live in a social world I'm either pinging them on Twitter or reaching out via email. Or my preference is if it's a blogger, the best way to test the water is to comment on their posts or tweet back/follow on Twitter." This approach helps build goodwill with an influencer in advance of starting a two-way dialogue, and it also prevents the influencer from wrongfully interpreting his attempts as "unsolicited sales outreach."

From there, Purcell waits to see what the influencer does next: "Do they encourage me to write back on their blog? Do they share more data with me? We have a program here at HPE where it's a blogger program [and] we have a sizable technology event twice a year called HPE Discover. We engage with bloggers throughout the year, and then we get a subset of those (the more influential ones) and invite them to come to the event. We also have things like Tech Days, where I'll invite a group of likeminded individuals on a particular topic and we spend a couple of days with them. To me it's all about relationship building and really understanding what they like and dislike. There's a value in the events because I get to meet people and build relationships and the influencers get exposed to the latest technology which is what they write about and share on their social channels – so it's a real win-win."

HOW to Reach Out

After determining who should manage outreach, and effectively engaging the influencer prior to outreach, it is finally time to directly contact the influencer. While social engagement is generally the best way to "warm up" an influencer, email is the best way in most countries to reach out to influencers with a proposed opportunity for collaboration.

For starters, let's look at what NOT to do:

✗ The email is way **too long**.
✗ It is **not a personalized message** tailored for the influencer.
✗ There is **too much about the brand** and what the brand needs.
✗ Not enough appreciation for the **influencer's limited time**.
✗ Unclear purpose—it should answer **what exactly the brand is looking for**.

With this in mind, here is how to structure an effective outreach email:

✓ **Strong Subject Line**: This determines whether your email gets opened in the first place. Therefore, you should think carefully before quickly writing the first few words that come to mind. A strong subject line makes the influencer curious to open the email to find out more.
✓ **Short, Sincere Greeting**: Grab the influencer's attention by demonstrating a genuine interest in their content and expertise (show them that this is a tailored email, and not something being spammed to a hundred other similar influencers).
✓ **Who You Are, W-A-Y-A-F**: Get to the point right away—who you are, and W-A-Y-A-F—"What Are You Asking For." These are the two primary pieces of information the influencer will want to know when they receive an email from you, as you have no preexisting relationship.

✓ **Closing: Call to Action**: If the influencer reads to the end of the outreach email, then it is highly likely that they will reply (either positively OR negatively). Make it as easy as possible for the recipient to reply "YES" by ending with a clear and easy-to-execute "Call to Action."

✓ **Email Signature with Links**: To avoid sending an email with too much information, have a short list of relevant links underneath the email signature. If the influencer is interested in learning more, they will have all the information readily available, but the email body itself will contain all the necessary components to help the influencer make a decision quickly about whether or not they're interested in the opportunity.

Don't immediately press "send"—wait for when the timing is right to avoid your email getting stuck in an inbox black hole. Monitor their primary social channel and get a sense of when they are active. Once the influencer posts new content or engages with someone else online, send the email and then give a quick "heads up" on their primary social media channel. Most influencers are happy to answer questions after they have a moment to review. One email may not be enough to elicit a response, so wait at least 24 hours and send a short follow-up message referring the original note. Do NOT spam influencers with multiple emails, and to gain mindshare between emails, continue to engage with their content online in a thoughtful manner (e.g., relevant comments, not "hey, did you get my email?") to serve as subtle reminders to the influencer that they should write back.

Keep in mind, email may not be the best way to get in touch with influencers in every circumstance. Mae Karwowsky, chief executive officer of Obviously, an influencer marketing technology company, did a campaign in Thailand, where few influencers had their email anywhere on their profile. This is very different from the USA, where most influencers have a dedicated email address on their social media profiles to make it easy for interested parties to get in touch. She explains, "We had to direct message the Thai influencers and come up with totally new ways of reaching out to people who we wanted to work with, but even then it generally took multiple messages and multiple online conversations." It is not just Asian influencers: "a lot of influencers in Europe will find my personal Instagram account and inquire if we are a legitimate firm. I even once had a father in Saudi Arabia call me to see if our outreach to his two daughters who were both beauty influencers was for real."

Getting Outreach Right: Keep It Focused, Keep It Personal

It's critically important for marketers to be patient and intentional in their outreach to influencers. By taking the time to learn more about what the influencer is known for among their audience, marketers will be able to better tailor their outreach approach to significantly increase the odds that the influencer will be open to exploring areas of collaboration. Don't start spamming with emails—that is a sure-fire way to never receive a reply or face public humiliation with a C.C. Chapman—Ragu-esque incident.

Looking ahead to Chapter 9, assuming a marketer's influencer outreach is successful, from this point he or she should consider the different ways that brands and influencers can collaborate. While there is an unlimited number of ways influencers and brands can work together to create mutual value for their shared target audience, the next chapter introduces a series of potential paths to take.

Marketer's Cheat Sheet

- **Be Patient**: It's very easy for marketers to get excited after they identify the perfect influencer and want to immediately reach out. A generic email sent without any prior online engagement will more often than not fall to the bottom of the influencer's inbox.
- **To Pay or Not to Pay**: Most of the time the decision to pay or not to pay is already made for the brand, because certain industries have to pay, while others cannot due to legal or ethical reasons.
- **Short-Term or Long-Term**: Paid engagements are most often short-term, one-off campaign collaborations between brands and influencers in the B2C space; while long-term influencer relations tend to be organic relationships in the B2B space or part of an "always on" B2C engagement approach where the influencer is paid upon "activation" when the right opportunity presents itself.
- **Clear Ownership**: Without establishing clear ownership of influencer outreach, the same influencer may be contacted by multiple individuals within the same brand. This can be extremely damaging, because in the eyes of the influencer whoever contacts them is "the brand." Make sure there is transparency and ownership around who reaches out to influencers on behalf of the brand.
- **Become Known First**: Prior to any direct outreach to an influencer about a specific opportunity to collaborate, the person responsible for outreach should invest time in engaging with the influencer's content and demonstrating a genuine interest in what they produce for their audience.
- **Personalize, Concise Outreach**: When it is finally time to reach out to the influencer directly, prepare a personalized email that demonstrates why the influencer is a perfect fit, and keep the message concise with a clear call to action. Don't just press "send"—observe their online activity and wait until the influencer is active and more likely to be checking email.

Notes

1. Chapman, C.C. "Ragu Hates Dads." *C.C. Chapman*, 2011, www.cc-chapman.com/2011/ragu-hates-dads/.
2. Chapman, C.C. "My Final Word on Ragu." *C.C. Chapman*, 2011, www.cc-chapman.com/2011/my-final-word-on-ragu/.
3. Schaefer, Mark. *The Rise of Influencer Marketing in B2B Technology*, 2017.
4. Solis, Brian. *Influence 2.0: The Future of Influencer Marketing*, 2017.

9

Working with Influencers: Potential Paths to Take

In 2016, when Marriott International set out to expand its Moxy Hotels brand from Europe to the USA, its marketing team knew they needed a creative approach to build brand awareness among American millennial consumers—Moxy's target audience. Video was decided upon as the ideal medium to communicate the brand's young and playful nature, making YouTube the natural platform of choice, given its position as the dominant video website in the US market.

They partnered with Taryn Southern, a YouTube personality and digital strategist with approximately 500 million subscribers, to help develop and host the video series they called "Do Not Disturb." Rather than pay for a short-term one-off video engagement, Moxy's brand management team opted for an ongoing video series to sustain audience engagement building up to the launch of the brand's first hotel opening in New Orleans later that year. In each episode, Southern interviews and gossips with fellow YouTube influencers in a Moxy hotel room, while both are dressed in their pajamas and casually reclining on a bed. Watching the videos gives viewers the feeling that they have been invited to a late-night slumber party with Southern and her guest, and things can get pretty goofy (Fig. 9.1).

Vicki Poulos, brand director for Moxy, explains, "We wanted to create a fun virtual extension of what the brand is all about…[and] it was important to find the right [influencers] for 'Do Not Disturb,' who would actually stay at a Moxy hotel.[1]" That is exactly what Moxy found in Southern and her assortment of young, talented content creator guests.

© The Author(s) 2018
J. Backaler, *Digital Influence*,
https://doi.org/10.1007/978-3-319-78396-3_9

Fig. 9.1 Taryn Southern and fellow YouTuber Bart Baker play "Truth and Dare"

Aside from Moxy branding on the pillows, the brand's formal presence in the videos is fairly minimal. Even behind the scenes, Southern receives a lot of freedom to develop the series based on her understanding of Moxy's target audience. "Normally, a brand will hire an influencer to just show up and tweet or Instagram about whatever the collaboration is about. In this case, Marriott hired me as a true partner, to produce the video series and develop the marketing campaign – I think the result is a much more authentic product," says Southern.

This direct partnership model is a real value for Marriott, because Southern plays multiple roles throughout the production process. Through tapping into Southern's production company, personal influencer network, and digital expertise, Marriott avoids having to deal with advertising agencies, other production companies, influencer agents, and multichannel networks—saving a lot of time or money that is often spent on these industry middlemen.

"The relationship has been really rewarding for me," Southern says, "because I feel that I'm not just a content producer for Marriott, but I'm also a strategist – we worked together every step of the way. It's beneficial for both parties to start their content strategy working hand-in-hand. They've been relatively hands-off, but also collaborative, leading with: 'how can we help you? How can we amplify what you're doing because you guys know how to do this stuff?' It's been a great partnership." Marriott's team felt the same way and went on to rehire Southern to continue the series with more videos.

Of course, as a large multinational company, it was not easy for Marriott to take such a hands-off approach to the campaign's production process—especially given some of the show's edgy and provocative subject matter. Marriott provided the online publication Digiday with a preview of the full series a month prior to the show going live on YouTube. When asked why it took so long for the first video to be published, David Beebe, vice president of global creative and content marketing at Marriott, told the publication,[2] "There's been a few more rounds of approval than usual for this one."

Ultimately, Marriott made the right decision for the Moxy brand and approved the video series. The collaboration was big success and Marriott hired Southern to produce a second season of "Do Not Disturb." Video collaboration is just one of many different ways brands can work with influencers to achieve their brand goals. This chapter introduces best practices for working with influencers, common business initiatives that benefit from influencer inclusion, and a series of influencer collaboration tactics for brands to consider when building their influencer programs.

Influencers Know Best: Provide "Freedom Within a Framework"

One of the greatest challenges brands face when working with influencers is giving them creative freedom in the production process. Brands are accustomed to working with expensive creative agencies to produce content, which positions the agency staff as order takers—whatever the brand wants, the agency needs to deliver (no matter how many rounds of revisions are requested, and no matter how frustrated the agency's team may be behind the scenes). As a professional services firm, the creative agency's number one goal is always: delight the client at all costs.

Influencer collaboration, however, is much different. Oftentimes, influencers are both the content creator (developing content that a creative agency would typically be hired to produce) and a distribution channel (they are the ultimate gatekeeper to their audience of target consumers for the brand). Oftentimes, brands overvalue the distribution side of the influencer's role and significantly underestimate the effort required for the influencer to produce content. This push and pull between influencers and brands throughout their collaboration can be one of the greatest points of tension (Fig. 9.2).

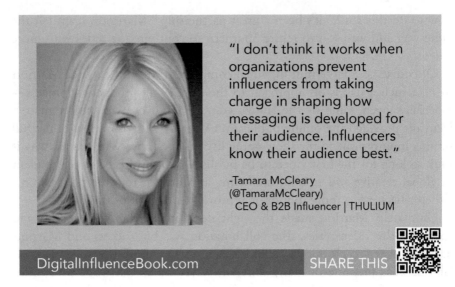

"I don't think it works when organizations prevent influencers from taking charge in shaping how messaging is developed for their audience. Influencers know their audience best."

-Tamara McCleary
(@TamaraMcCleary)
CEO & B2B Influencer | THULIUM

DigitalInfluenceBook.com   SHARE THIS

Fig. 9.2 Tamara McCleary

Brittany Hennessy, senior director of influencer strategy and talent partnerships at Hearst Magazines Digital Media, once worked on an influencer campaign in which the brand wanted an influencer to say specific words and to mention the full name of the product in every shot. "They were going to pay this influencer a lot of money but were treating her content like an ad. We guided them away from that approach because it undermined what makes influencer marketing so successful. Making her sound like a spokesperson instead of the influencer they know and love is the quickest way to make her audience suspicious of the content she's promoting."

She explains how frustrating the working relationship can be for brands that are accustomed to order taker creative agencies: "Many brands think, 'we're paying these people, why can't we tell them what to do?' But the reality is, brands are working with creators because creators already know what to do. As a brand, you're directing the conversation to make sure content is published on your timeline and includes certain talking points, but you're not making an advertisement and it can be difficult to remember that."

This idea of the brand serving in a director role, and not dictating every aspect of the influencer's content production process, is what Tom Doctoroff calls providing "freedom within a framework." Doctoroff, who is a senior partner at the branding agency Prophet, feels strongly that "brands have to give influencers a template, a framework for participation."

This template can be higher level based on mutually agreed upon collaboration briefs, or a literal template content producers can use to ensure items like product logos and other tactical elements fit the brand's standards. "It really comes down to making sure that brands are framing the debate but leaving enough space for influencers to participate naturally," he says.

Hennessey and Doctoroff are not alone in this belief. Lee Odden, a prominent B2B content marketing influencer and chief executive officer of TopRank Marketing, feels the same way: "Letting go is one of the greatest challenges brands face when it comes to influencer marketing. When it all comes down to it, the value of the influencer and their network originates from the influencer's own point of view and approach."

This is not to say brands should just let influencers have 100% complete control of the content production process, but they need to be willing to let go "and trust that the influencer (with guidance) will do what's best to engage their audience, otherwise brands will miss out on the influencer's intimate knowledge of what its audience actually wants."

Research conducted by global public relations firm Edelman reaches the same conclusion:

[Influencers] must remain **genuine and relatable,** which makes working with them a challenge as brands must learn to collaborate vs. dictate heavy handed marketing. Also, brands must develop repeatable ways they can work with all levels of these types of influencers. As it is an emerging space often requiring complex contracts, disclaimers and transparency – it brings new operational dynamics to the table.[3]

Not everyone takes the influencers' side in this debate though. Reb Carlson, associate account director at the digital agency Wunderman, feels brands are right to believe they should be able to dictate what influencers do. "If influencers are looking to collaborate with brands, they need to follow the brand brief. If there is a formal contract, then they are a formal contractor of that company" (Fig. 9.3).

Carlson has seen a lot of influencers view the client brief more as a set of general guidelines rather than something that needs to be strictly adhered to. "The reality is, if influencers are being paid by a brand, then they need to work within the framework of the brief. I think that a lot of influencers need to understand that it's still business and they need to deliver the work the client wants just like any advertising or creative agencies has to as well."

Overwhelmingly, across the industry professionals and influencers interviewed for this book, "freedom within a framework" wins out as the best

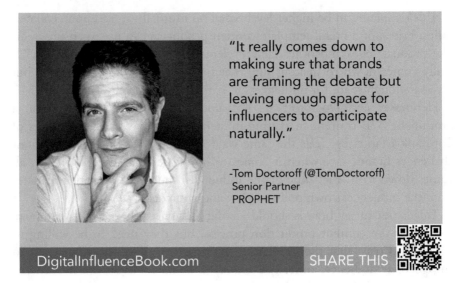

"It really comes down to making sure that brands are framing the debate but leaving enough space for influencers to participate naturally."

-Tom Doctoroff (@TomDoctoroff)
Senior Partner
PROPHET

DigitalInfluenceBook.com   SHARE THIS

Fig. 9.3 Tom Doctoroff

practice for working with influencers. Without a doubt, Marriott would not have achieved such a high level of success launching its Moxy hotel brand in the USA if their team had taken a heavy-hand to Southern's production process. Instead, by providing Southern with the necessary resources, guidance, and direction throughout their collaboration, Southern was able to deliver a final product that both achieved the brand's goals while being perceived as genuine and relatable by Southern's and her guest influencer's fan bases.

Realistically, a brand is not going to come initially and create a script for an influencer to follow that will resonate with their fans. The influencer knows how to do it, and they have proven they can do it. Brands need to empower influencers to create a product that aligns with their audience and not be overly restrictive, and providing the framework to guide this process is key.

When to Engage: Common Business Initiatives to Partner with Influencers

As brands explore more ways to incorporate influencers into their businesses, both sides are learning. Learning to provide freedom within a framework is one aspect, but brands are also figuring out the best ways to incorporate influencer marketing into wide-ranging corporate initiatives. As influencer marketing gains traction, and more areas of the organization become receptive to including influencers in their traditional business practices, there are a variety of areas where influencers can have an impact. Actual business areas vary from company to company, and the only real limit is a brand's creativity, but the following are a series of business areas where influencers have proven to be successful.

Corporate Reputation Management

One established practice for brands to engage influencers is to work with them as part of proactive corporate reputation management. As companies identify high-risk trends that could negatively impact their industry or firm's reputation, the right influencer strategy can help get ahead of potentially damaging PR. For example, in 2015 *The New York Times* reported that Coca-Cola engaged influential scientists to shift the health conversation away from people needing to cut out sugary drinks to needing to get more exercise instead. Steven Blair, an influential exercise scientist who was part of Coca-Cola's efforts, was publicly quoted as saying, "Most of the focus in the popular media and in the scientific press is, 'Oh they're eating too much, eating too much, eating too much' – blaming fast food, blaming sugary drinks and so on… And there's really virtually no compelling evidence that that, in fact, is the cause.[4]"

In the automotive industry, General Motors (GM) worked with Celebrity Influencers to shift negative perceptions that the firm only produced gas-guzzling SUVs to a more positive, eco-conscious narrative. In 2003, Arianna Huffington launched the Detroit Project,[5] a group which produced a series of provocative commercials connecting consumers' purchase of SUVs, and the extra gas required to fuel them, to indirectly funding terrorist activities in the Middle East. GM's Chevrolet Suburban was a prime target, which quickly led to the firm being demonized in the press.

GM's vice president of global communications, Tom Kowaleski, hired Charlie Windisch-Graetz, managing partner of C4 Consulting, to develop an influencer-led strategy to increase awareness of the firm's $1 billion+ investments in hydrogen-powered cars. Windisch-Graetz convinced GM to fly in its $35-million-dollar prototype from Geneva, and he and his team built a test track at the Santa Monica Airport in Los Angeles. They sent exclusive test drive invitations to Celebrity Influencers known for their focus on the environment, such as Cameron Diaz and Ed Begley Jr. They were quick to accept the invite to be the first civilians ever to drive a hydrogen-powered car. "By the end of the event, and through the subsequent media attention, we were able to shift the discussion from 'GM is building Suburbans' to 'GM is spending over a billion dollars to develop the future, environmentally-friendly automobile,'" explains Windisch-Graetz.

Amplification of New Product Launches and Brand Events

After developing new products, beauty companies generally go on to engage influencers for their product launches to tap into active online followings to make a big splash when the product goes live. Influencers also play a central role in amplifying product launches and events for a wide spectrum of other industries. Outside of the beauty industry, the online gaming industry regularly releases new products as well.

Electronic Arts is one of the biggest players in this space, and the company has shifted its product launch advertising spending away from traditional TV ads to influencer marketing. On a 2017 investor call, EA's chief financial officer, Blake Jorgensen, explains their rationale: "When we see something that works, we continue to fuel it. We're driving more and more of our marketing away from traditional media and into the influencers of the world. And because of that, you're going to see different patterns in marketing spend.[6]"

In 2016, Lenovo introduced a new category of mobile products: the Moto Z Smartphone and detachable accessories called Moto Mods. To drive consideration among target consumers and to educate their audience on ways to use the Moto Mods, Lenovo partnered with more than a dozen up-and-coming influencers to bring to life fun use-cases that fit in their personality and lifestyle.

One of those influencers was the world-renowned Chinese fashion designer Vivienne Tam, who is known for her vibrant, eclectic East-meets-West aesthetic. Lenovo incorporated the Moto Z and Moto Mods into her New York Fashion Week runway show in September 2016 in unique and unexpected ways. While front-row guests used the Hasselblad True Zoom Mod to snap photographs, models walked the runway sporting Moto Z phones with limited-edition style shells, featuring prints which matched Tam's Spring 2017 collection.

Working directly with Tam and her team in the months leading up to NYFW, Lenovo collaborated on all aspects of the exclusive style shell design from the print selection using the designers yet-to-be-released fabrics to the development of the innovative 3D textured printing on the case itself and the incorporation of both brand logos. Tam also produced a special embellished Moto x Vivienne Tam clutch bag to match one of the shells and house gifted Moto Z phones and Moto Mods for VIP guests.

Product Development Inspiration

While many brands still view influencers as ways to get information out, some of greatest benefits happen when brands tap into influencers' unique perspectives as part of their internal product development process. Simon Kemp, chief executive officer of the marketing firm Kepios, observes: "That works really well because you have an outside interface [the influencer] that allows you to get genuine public feedback into product development." He feels that an influencer, when used properly, behaves as a third-party interpreter: "they're not necessarily acting in the best interests of the brand, they're acting in the best interest of the thing that they care about and if the brand does its job properly it should be listening to that, it should be learning and catering to that influencer."

In addition to incorporating their feedback on existing products, many companies work with influencers to co-develop products from their inception. This is especially the case for the cosmetics industry, where studies have found influencer collaborations lead to an average of two times the dollar volume of traditional celebrity collaborations, and 92% of makeup users get information about new products from these influencers.[7]

Capitalizing on this trend, Sephora collaborated with Huda Kattan, an established beauty Category Influencer with more than 20 million followers on Instagram and with a strong following in the Middle East. Shortly after launching their co-developed Huda Beauty line, it quickly became a top seller throughout Sephora's local retail outlets in the United Arab Emirates.[8]

Similarly, MAC Cosmetics enlisted 10 beauty influencers from eight countries (USA, UK, Canada, Brazil, Middle East, France, Germany, Australia) to co-develop a new line of lipsticks for each of their country's local consumer markets. The group was invited to MAC's development laboratory in Toronto, where each influencer created their own lipstick through testing and feedback sessions with MAC's product development team. Catherine Dougherty, a senior vice president at MAC, told the website PopSugar, "They are really well-versed on the products, the textures, the finishes, the ingredients, the undertones…I felt like when we were working with these people that some of them know our product better than we do sometimes.[9]"

By incorporating influencers at the product development level, brands can gain valuable insights into what their customers really want. The right influencers are able to interpret their understanding of their audience's preferences and help feed that into a cosmetic brand—or any brand for that matter—new product roadmap.

Product Reviews When Entering New Markets

When brands enter a new market (either a new product category or new geography), third-party validation from relevant, authoritative insiders is an essential aspect of product positioning. Reviews can make the difference between acquiring a new customer and missing out on their business, especially for products and services sold predominantly online. This was definitely the case for sportswear company Reebok, when it planned to enter the premium running shoe market with its $150 a pair Floatride Run model.

Up until the release of Floatride, running shoes had not been a product category of focus for the brand, making Reebok far from a runner's first choice when shopping for new high-end running sneakers. Instead of televised ads featuring famous running athletes, Reebok chose to focus on engaging a large number of Micro-Influencers to build up product reviews on its online store.

It collaborated with Experticity, a retail marketing agency, to provide a 1200-member group of Micro-Influencers (including sneaker salespeople and running club instructors) with free samples of the Floatride shoe to try for themselves. Reebok then directed the participating Micro-Influencers to leave reviews sharing how much they liked (or disliked) the product. The Micro-Influencers were objective reviewers; outside of providing the free sneakers, Reebok did not offer any monetary incentive for their participation.

The results were impressive. By the time the Floatride shoe went to market, the sneaker model acquired more than 600 product reviews—a 50% product trial-to-review. David Pike, associate manager of brand management for Reebok, told Digiday, "Influencer marketing is hard to quantify, but product reviews are tangible. When people see customer reviews for a new product, they feel more informed and confident to buy.[10]"

Search Engine Optimization

With search engine algorithms constantly evolving to prevent unscrupulous Internet marketers from taking advantage of the system to gain visibility on search engine results pages, one source of enduring SEO credibility is back-links from authoritative websites. However, it can be difficult for brands to convince these established websites to link back to their company website and product pages. This is where influencers, who often maintain their own authoritative websites, can be extremely beneficial. For example, a link back from an influencer's blog may deliver a short-term traffic boost from blog's visitors, but the greatest impact will result from the improved search results ranking due to the influencer's website link "telling" the search engine that the brand's website is high-quality.

The UK travel company Thomas Cook Airlines regularly engages influencers to develop content that links back to their website. Diego Puglisi, search marketing manager at Thomas Cook, told Econsultancy about how changes in SEO rank is a key factor in measuring the results of influencer collaboration: "We measure rank changes from an SEO point of view as well as direct traffic to the site from the influencer's content.[11]" By having travel influencers link to Thomas Cook's website, the airline is able to boost its visibility among prospective travelers authentically, without needing to hire a third-party firm that may use dishonest link-building strategies that could damage the Thomas Cook brand.

How to Engage: Different Paths to Influencer Collaboration

In incorporating influencers into business initiatives like corporate reputation management, product launches, and product development, brands can choose from a wide selection of more tactical activities to collaborate on with influencers. In an interview with Adobe's CMO.com editors, TopRank CEO Lee Odden sums it up well: "Influencer relationships are a valuable resource for a brand. Those relationships are an asset and, from a marketing standpoint, you can tap those relationships to help create content for campaigns, projects, or even ongoing advocacy programs. At the same time, if you're a PR person, you can tap those very same influencers as subject matter experts to contribute articles for a respected industry publication. There are multiple ways to integrate influencers across functions, but they must be managed properly.[12]"

The ways brands can integrate influencers into their businesses are ever-changing as next generation technologies come online, and as influencers develop new ways to engage their audiences. At the time of writing this book, the following examples of influencer collaboration are regularly used by industry practitioners. This is far from a comprehensive list, but these examples serve as a valuable resource for brands as they consider different ways to work with influencers in their own industry.

Incentivize Influencers to Share Brand Content

Perhaps the most commonly used form of influencer collaboration is incentivizing influencers to create and share brand content. In a B2C context, this could mean Reebok in the Floatride example, providing Micro-Influencers with free sneakers to incentivize them to share photographs of the new running shoe line on prominent social media sites like Instagram, Twitter, and Pinterest. In a B2B context, it could be facilitating exclusive access to senior executives for interviews that the influencer can then publish on their personal blog or video channel, as was the case with the American technology firm Teradata.

Co-create New Content with Influencers

Beyond incentivizing influencers to share content directly on their own personal platforms, brands can co-create content with influencers and share it among both the brand's and the influencer's audiences, giving each exposure to new eyeballs and a chance to grow their respective online followings. On the B2C side, L'Oreal has done this well through its unbranded fashion website, FAB Beauty, where it features the content and stories of beauty industry trendsetters. Similarly, in the B2B space, Adobe, an American software company, operates a website CMO.com, where it features interviews with Category Influencers and other influencer-generated content. Other forms of co-created content to consider include webinars, videos, and podcast interviews.

Cultivate Brand-Managed Influencer Communities

An extension of co-creation, brands can also develop their own online communities where influencers share content related to the brand. The Coca-Cola Company did this when it launched "The Coca-Cola Journey," where it transformed its homepage into a content hub the likes of the Huffington Post. According to Ruben Ochoa, an influencer marketing strategist who was part of the launch, "Coke initially hired an editorial staff, but it wasn't enough because they didn't have the subject matter experts. I ended up running a 25-person influencer program with them to help Coke expand into various content categories like 'women's lifestyle, 'food', 'travel', 'innovation' – it's an excellent example of an influencer contributor network that helps fill a large brand's needs."

Repurpose Existing Influencer Content

Most content has a much longer shelf-life than what it is initially created for. Brands should come up with creative ways to repurpose influencer-generated content. For example, if an influencer did a photo shoot for a one-off brand collaboration on Instagram, some of those photographs could be repurposed as photographs for digital or traditional print ads. DSW, the American footwear retailer, partners with its agency to incorporate influencer-generated content into online ads, html email images, and other aspects of its content marketing efforts.[13] By doing this, brands can also save big money by avoiding the need to hire an expensive celebrity for an all-new photo shoot.

Hold In-Person Events

Invite influencers to events that are being organized by the brand or industry events where the brand is a major partner. This is the approach the Chinese technology firm Huawei uses to engage B2B influencers at its annual conferences, as well as major industry events like the Mobile World Congress in Barcelona. In addition to serving as a way for influencers to experience the brand in person, by incorporating the right Category Influencers brands can benefit from the additional social amplification of having these influencers share live at the event. This is especially true for B2C companies that regularly incorporate influencers into product launch events as a way to help get the word out or to create a unique experience for their customers. American Express did this by developing an influencer advisory board that it calls The Platinum Collective. The group of 15 luxury Category Influencers help the brand develop event experiences and other concepts to benefit their high-end platinum credit card holders.[14]

Develop Influencer-Driven Affiliate Sales Programs

One of the most mutually beneficial forms of influencer collaboration is via an affiliate partner program, in which a brand pays an influencer a percentage of sales that originate from the influencer's promotional efforts. In the

B2C space, brands typically provide influencers with special affiliate links that can be used to track sales. The American consumer goods company Kimberly-Clark regularly provides its influencers with special coupon codes to attribute the sales they help generate. Implementing an effective influencer affiliate sales program may be more difficult for B2B brands, given the higher price point of their products, and generally longer sales cycles. As attractive as affiliate programs are to generate new sales, brands should proceed with caution and pay careful attention to the suggestions in Chapter 10: *Know the Risks: The Dark Side of Influencer Collaboration* in order to avoid potential costly legal mistakes.

Run Competitions and Giveaways

A great way to drive brand engagement among influencers' followings is by partnering with influencers on competitions and giveaways. Brands provide influencers with an exclusive set of in-demand products or prizes, and the influencers' audience participates in a competition to get their chance to win. Zevia soda partnered with a range of Category Influencers including Jade Sheldon, a Category Influencer who is known for cooking and photography, on a giveaway where participants had to tag three friends and use a special hashtag for a chance to win a $500 Visa gift card. Through her audience's participation, the soft drink brand built brand awareness and identified new potential consumers to target their future marketing efforts.

Experiment with Influencer Social Media Account Takeovers

Similar to giveaways, influencer social media account takeovers is another creative way for brands to reach new audiences, increase engagement, drive Web traffic, and build brand awareness. They are not for everyone; giving an external influencer control of a brand's social media account like Instagram or Twitter requires a lot of trust between both parties, but some brands have run successful experiments to date. The financial services firm Capital One collaborated with three photography Category Influencers over the course of five weeks, giving them access to the brand's Instagram account. They were directed to post photographs of interesting things they keep in their wallet as play on the brand's famous slogan, "What's in your wallet?" The brand benefited by gaining exposure to the influencers' audiences, that it could not have achieved through its own organic Instagram activity.

Influencer Engagement in Two Words: Freedom and Creativity

There are an unlimited number of ways for brands to collaborate with influencers and incorporate them into their various business initiatives like corporate reputation management, product development, and product launches; it is all about how creative brands and influencers are in coming up with novel ways to join forces to engage their mutual target audience. However, providing influencers with the necessary freedom to authentically engage their fans throughout the collaboration is essential for success. Offering "freedom within a framework" is what ensures influencers maintain this authentic connection, while still fulfilling the brand's needs and requirements.

This was certainly the case for Marriott when they hired Taryn Southern to develop their "Do Not Disturb" YouTube series. Both brand and influencer collaborated from the very inception of the idea to come up with a creative way to build brand awareness of Moxy hotels by tapping into the audiences of Southern and her highly influential guests. By providing Southern freedom to create content that she knew would resonate with her following, the series attracted the right audience that ultimately went on to become many of the first guests at Moxy's new hotels in the USA.

Influencer collaboration can be highly beneficial for brands, but it is not easy to put in practice what often seems so straightforward on paper. There are many risks that brands need to be aware of as they move forward in building their influencer programs. There is a very real dark side, full of legal reputational risks for brands, and widespread ways that influencers are gaming the current environment to create "artificial influence." Both sides are addressed in the next chapter, *Know the Risks: The Dark Side of Influencer Collaboration*, which is not to be missed.

Marketer's Cheat Sheet

- **Freedom Within a Framework**: It is reasonable for marketers to want to control the entire creative process when collaborating with influencers. However, the entire reason brands work with influencers is that influencers know what their target audience wants. Therefore, marketers should act as directors, not as dictators, to ensure a successful collaboration like Marriott's work with Taryn Southern.
- **Ensure Authenticity**: An influencer agreeing to collaborate is one thing, but engaging their audience in the right way to benefit the brand is another. Marketers need to be sure that what they ask of the influencer does not go

against what their audience expects of him or her. Take Lee Odden's advice from this chapter: "The value of the influencer and their network originates from the influencer's own point of view and approach." Otherwise, if the influencer does not connect to their audience, then the whole collaboration becomes a waste of time and money.

- **Be Creative**: There is no limit to how marketers can work with influencers both at the campaign level and across broader, more strategic corporate initiatives. Invite influencers to help with product development or create an advisory board to help improve your customer experience. As technology continues to advance, influencers will develop even more ways to engage their audiences that are unthinkable today, so brands must be ready to adapt. Consider this—a brand could invite an influencer to keynote a conference "live" via hologram. It is not as far away as one may think.
- **A-B-E (Always Be Experimenting)**: While incorporating influencers into events may work for a brand today, they may find that other forms of collaboration, like giveaways or account takeovers, may be more effective tomorrow. The best thing marketers can do is to continue to experiment with different forms of collaboration that push the limits and stay fresh/novel with their audience. This even means B2B marketers adapting "B2C influencer engagement models" to make them work in a B2B context.

Notes

1. Oates, Greg. "Moxy Hotels' New YouTube Series Is Not Intended for Mature Audiences." *Digiday*, 13 Nov. 2015, digiday.com/marketing/moxy-hotels-new-youtube-series-not-intended-mature-audiences/.

2. Oates, Greg. "Moxy Hotels' New YouTube Series Is Not Intended for Mature Audiences." *Digiday*, 13 Nov. 2015, digiday.com/marketing/moxy-hotels-new-youtube-series-not-intended-mature-audiences/.

3. Armano, David. "The Death of Content Marketing." *Edelman*, 5 May 2016, www.edelman.com/post/death-of-content-marketing-why-brands-must-become-cultural-currency/.

4. O'Connor, Anahad. "Coca-Cola Funds Scientists Who Shift Blame for Obesity Away from Bad Diets." *The New York Times*, 9 Aug. 2015, well.blogs.nytimes.com/2015/08/09/coca-cola-funds-scientists-who-shift-blame-for-obesity-away-from-bad-diets/?_r=0.

5. Walker, Rob. "What Would Arianna Drive?" *Slate Magazine*, 13 Jan. 2003, www.slate.com/articles/business/ad_report_card/2003/01/what_would_arianna_drive.html.

6. Grubb, Jeff. "Electronic Arts Is Spending Even More on 'Influencers'." *VentureBeat*, 27 July 2017, venturebeat.com/2017/07/27/electronic-arts-is-spending-even-more-on-influencers/.

7. Gonzalez, Amanda. "Do Consumers Crave Collaborations?" *The NPD Group*, 1 Nov. 2016, www.npd.com/wps/portal/npd/us/blog/2016/do-consumers-crave-collaborations.

8. Hamdan, Lubna. "The Business of Blogging in Dubai and the GCC." *ArabianBusiness.com*, 16 Apr. 2017, www.arabianbusiness.com/the-business-of-blogging-in-dubai-and-gcc-670973.html.

9. Orofino, Emily. "10 Influencers Just Designed New MAC Lipsticks—And They're Not What You'd Expect." *POPSUGAR Beauty*, 1 Apr. 2017, www.popsugar.com/beauty/MAC-Cosmetics-Beauty-Influencer-Collection-Interview-2017-43332924.

10. Chen, Yuyu. "How Reebok Used Influencer Reviews to Break into the Competitive Running Category." *Digiday*, 20 June 2017, digiday.com/marketing/reebok-used-influencer-reviews-break-competitive-running-category/.

11. Gilliland, Nikki. "How Influencers Can Impact SEO: Q&A with Thomas Cook Airlines." *Econsultancy*, 2 Mar. 2017, econsultancy.com/blog/68856-how-influencers-can-impact-seo-q-a-with-thomas-cook-airlines/.

12. Rumsey, Angela. "Lee Odden Of TopRank Marketing Influences Through Influencers."*CMO.com by Adobe*, 12 July 2016, www.cmo.com/interviews/articles/2016/7/11/lee-odden-of-toprank-marketing-influences-through-influencers.html#gs.ZULA4Ls.

13. Pathak, Shareen. "Why Shoe Retailer DSW Is Turning Influencer Content into Ads." *Digiday*, 13 Oct. 2017, digiday.com/marketing/shoe-retailer-dsw-turning-influencer-content-ads/.

14. Chabbott, Sophia. "Amex Taps Jennifer Fisher and More Social Influencers for New Initiative." *WWD*, 3 Nov. 2016, wwd.com/business-news/marketing-promotion/amex-platinum-collective-jennifer-fisher-10698512/.

10

Know the Risks: The Dark Side of Influencer Collaboration

It is March 2014, and The Walt Disney Company has just agreed to acquire Maker Studios—the largest multi-channel network (MCN) in the world, with 5.5 billion views a month from a subscriber base of 380 million[1]—at a price tag of $500 million. The deal seems like a match made in heaven, combining Disney's stable of characters and intellectual property with Maker's distribution and programming expertise.

MCNs help monetize video content on online video platforms, like YouTube, and "offer services that may include audience development, content programming, creator collaborations, digital rights management, monetization, and/or sales."[2] With approximately 55,000 YouTube channels under management, Maker is poised to help Disney reach young consumers around the world.[3]

Maker's top channel under management is "PewDiePie," produced by a 24-year-old Swede named Felix Kjellberg. The Category Influencer produces content primarily on video games, with some comedy thrown in from time to time. Through the acquisition, the Disney brand is now associated with all of the YouTubers under Maker's management, including Kjellberg.

Fast-forward to three years later, where Kjellberg (now 27) has achieved new levels of fame—with 53 million subscribers, he is YouTube's biggest star. *Forbes* reports that he earned $15 million in 2016 alone, in addition to publishing his first book, *This Book Loves You*, and signing on for a reality series called "Scare PewDiePie" on YouTube Red, a paid streaming service. Despite his high profile, Kjellberg continues to push the envelope with his content and begins experimenting with more extreme ways to increase video views.

© The Author(s) 2018
J. Backaler, *Digital Influence*,
https://doi.org/10.1007/978-3-319-78396-3_10

In January 2017, he uses the freelancer marketplace Fiverr to hire someone willing to submit a video of themselves dancing and holding up a sign of the client's choosing—and what he chooses to have the freelancer write on the sign shocks executives at Maker and its parent company, Disney. Publicized in the mainstream media by the *Wall Street Journal*[4] nearly a month later on February 14, the world finds out that Kjellberg hired Indian freelancers on Fiverr to dance around half-naked while holding a poster that reads "Death to all Jews."

Once the anti-Semitic video is brought to their attention, Maker immediately cuts ties with Kjellberg and provides the media outlet *Variety*[5] with the following official statement: "Although Felix has created a following by being provocative and irreverent, he clearly went too far in this case and the resulting videos are inappropriate. Maker Studios has made the decision to end our affiliation with him going forward."

But how do Kjellberg's actions reflect on the family-oriented Disney brand? What is Disney Chief Executive Officer Bob Iger's reaction, especially since he himself is Jewish? The fallout from Kjellberg's irresponsible and morally questionable behavior has left Disney to reconsider how it plans to integrate Maker into the Disney portfolio. Later the same month that the *Wall Street Journal* story breaks, Disney announces that Maker's network of content creators would be reduced from tens of thousands to a mere 300, and Maker's role going forward would be significantly scaled down, primarily serving as the digital and social marketing arm for its Consumer Products and Interactive Media division.[6]

Disney's $500 million MCN bet did not work out as planned because Kjellberg's actions forced the company to realize that there was too much brand and reputational risk at stake if it continued to associate itself with a loosely managed network of thousands of individuals that it has little control of. The same goes for any brand that works with influencers; there is always a risk that an influencer's unsavory actions will come back and tarnish the brand. This is one of many pitfalls marketers need to watch out for, as we explore the dark side of influencer collaboration.

Welcome to the Dark Side: What Risks Brands Need to Be Aware of

When it comes to influencer marketing, brands need to be aware of potential reputational and legal risks, such as what happens when an influencer's behavior goes against the values of the collaborating brand, as in the case with Disney and Kjellberg. Additionally, there are a series of considerations that brands need to watch out for when it comes to deciding whether to work with an influencer in the first place. Chapters 3 and 7 clearly make the point that looking at follower count alone is not a good measure to assess whether someone is influential, and this chapter goes into specifically why this is the case, exposing the various ways individuals can "buy influence" from their number of followers to their audience engagement rates. There is an entire sub-industry of service providers out there willing to help influencers game the system in order to catch the attention of brands willing to pay them lucrative fees in exchange for their supposed influence. There is no doubt that influencers can bring tremendous value to brands; however, brands should not be naive—there are a lot of risks out there, and it is important to be aware of them now to avoid costly mistakes in the future.

Guilty by Brand Association

Think back to the Italian fashion influencer Chiara Ferragni and Chinese cosmetics influencer Melilim Fu from earlier in this book—the goal of both individuals is not to partner indefinitely with established brands, but rather leverage their influence to launch their own branded product lines. This ties into Taryn Southern's advice from Chapter 5: "When you think about it, an influencer is a brand – a personal brand. Therefore, when a corporate brand wants to collaborate with an influencer, it's really two brands trying to come together." The challenge for corporate brands is that sometimes the influencer's personal brand can clash with the corporate brand's values and everything it stands for (Fig. 10.1).

The Disney-PewDiePie case is just one example of how an influencer's personal actions can reflect poorly on a brand. Beauty brand L'Oréal shared a similar experience when it partnered with British social activist and trans-gender model Munroe Bergdorf. Within 48 hours of announcing Bergdorf's appointment to L'Oréal's "Beauty Squad," a diverse group of Global Influencers who appear in the brand's ads and co-create content, the beauty brand fired her. Why? Because Bergdorf wrote a Facebook post addressing racism and white supremacy that led the *Daily Mail* to claim that she said all white people are racist.

The company released a statement over Twitter: "L'Oréal champions diversity. Comments by Munroe Bergdorf are at odds with our values and so we have decided to end our partnership with her."[7] Bergdorf claimed that the post was taken out of context, and that her being fired goes against

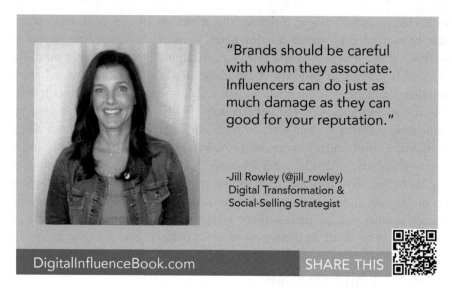

"Brands should be careful with whom they associate. Influencers can do just as much damage as they can good for your reputation."

-Jill Rowley (@jill_rowley)
Digital Transformation &
Social-Selling Strategist

DigitalInfluenceBook.com   SHARE THIS

Fig. 10.1 Jill Rowley

L'Oréal's emphasis on diversity, but the *Daily Mail* article was enough to make the brand reconsider their brand association.

B2B brands are not immune to potential brand damage from associating with the wrong influencer, either. Robert Scoble, a high-profile technology Category Influencer, leveraged his influence to become a Microsoft "Evangelist" and later a Rackspace "Futurist." While being associated with Rackspace, an American cloud computing company, Scoble attended an outdoor technology conference where attendees slept in tents and stayed up late drinking by campfires. According to a detailed account[8] by Quinn Norton, a journalist who was also attending the event, Scoble sexually assaulted her that night.

She recounts: "For years when I saw Scoble's name on something, or the mention of Rackspace, his employer, I flinched. I stopped consuming media that was supported by Rackspace, not wanting to feel those hands on me every time they were mentioned."

When Quinn went public, several other women also came out to share their similar experiences with Scoble, and all the technology firms Scoble had been associated with immediately turned on him, as their brands were tarnished in the process for supporting the advancement of Scoble's career while his inappropriate actions had been an open secret within the tech community for many years.

Tom Doctoroff, senior partner at the branding firm Prophet, explains, "There is a risk that influencer marketing (unless it is managed very carefully) can go wrong—and by go wrong I mean not be consistent with the what the message of the brand needs to be." He feels the more people who have a voice when it comes to how the brand is communicated in the marketplace, the more chaos there can be. "And when you have chaos—it's confusing for consumers and ultimately bad for the brand. It can either be off-brand or just disjointed or derogatory for that matter. It's the marketer's responsibility to harness the voices."

As part of the influencer identification process, brands should take the time to dig deep into potential influencer's background, especially for high-profile, ongoing brand associations. While technology platforms may be helpful to identify an influencer, it is important to not only speak to the influencer directly, but also to gain insight from others with knowledge of the individual's professional and personal background. There are no shortcuts to make sure an influencer is not only a good fit for the brand today, but also has a high likelihood of remaining a good fit (by not tarnishing the brand) in the future. Lastly, have the right contracts in place. "I think a lot of what Hollywood does with contracts needs to be applied to influencer marketing to ensure brands have the right protections in place," advises Cynthia Johnson, chief executive officer at the personal branding firm Bell + Ivy.

FTC and Other Legal Risks to Watch Out for

What happens when a brand's target audience is not able to distinguish sponsored content from content an influencer genuinely enjoys producing? This is the question a group of industry watchdogs are forcing the US Federal Trade Commission (FTC) to answer as they seek stricter guidelines on brands using influencer marketing to target children. The Center for Digital Democracy, Campaign for a Commercial-Free Childhood, and Public Citizen filed an official complaint with the FTC for what they consider "unfair and deceptive practice of targeting influencer marketing toward children," since children are too young to differentiate between sponsored and unsponsored content,[9] even when the influencer uses the appropriate ad disclosures. This is just one example of how government consumer protection agencies around the world are being forced to reexamine how they classify influencer marketing, and the various rules and regulations they have in place to ensure their nation's consumers are not taken advantage of.

The USA is far from the only country where regulators are keeping a closer eye on the development of influencer marketing. In the UK, the Advertising Standards Authority (ASA), UK's independent regulator of advertising across all media, has released guidance for influencers on disclosures. Similar organizations around the world that have set comparable measures include the European Advertising Standards Alliance, China's State Administration for Industry and Commerce, Brazil's Consumer Protection Code, and the ASA of Singapore. The ultimate aim of all of these governmental bodies is to make sure that influencer marketing is more transparent, so its citizens are not deceived into buying a product or service because they mistake a sponsored ad for a genuine influencer endorsement.

> "Think of an influencer's social media disclosure as a politician saying, 'I approve this message' in their political commercials."
> - Alexander H. Hennessy J.D.
> Co-founder, CREATORSCOLLECTIVE

As one of the most established markets for brand–influencer collaboration, the USA is a good benchmark to monitor as the FTC releases more guidance on proper disclosure. According to American legal experts, the FTC is unlikely to enact more restrictive measures, but rather opt to provide more detailed guidance. Ashley R. Villa, Esq., founder and CEO of Rare Global, a digital focused management company, explains, "We will learn in more detail how to implement these visual and verbal disclosures. As of now the FTC hasn't given too much real direction outside of the page online that talks in broad strokes."

Villa believes that the FTC will provide greater clarity, as right now there are a lot of open questions about how influencers can comply: "What does 'visual and verbal disclosure' really mean? Is that placing #ad as the first word of the copy of an Instagram photo? Can the #ad be the third line? On YouTube, where should the on screen visual disclosure lie? Burned into the video? Placed in the description box?" Regulators can do more to help clarify what best practice is for disclosure under different circumstances.

While it's commonplace for influencers in the USA and Europe to disclose brand affiliations, it's not necessarily the case in other markets. Soukaina Aboudou, a Middle East-based beauty influencer explains:

"Influencers here are still not very comfortable letting their community know when content is sponsored, so the overwhelming majority of them don't mention it." Aboudou recognizes that not disclosing is illegal, but she feels viewers in the Arab-speaking world don't tolerate brand collaborations yet and still consider them to be an abuse of the confidence they've put in the influencer. "This is the biggest problem that we have, and I'm personally determined to let my community know that sponsorships are not evil (it doesn't mean that we are lying about the products) we simply should get paid for all the effort and work we put into create content."

Even if an influencer fails to properly disclose a sponsored post, at the time of writing this book, they are unlikely to face serious legal consequences. Most issues flagged by regulators are resolved by simply having the influencer remove or update their post. Mondelez International, maker of the Oreo cookie brand, paid British YouTubers Phil Lester and Dan Howell to participate in its "Lick Race" challenge. The UK's ASA concluded[10] that the video postings did not properly disclose the paid association between the influencers and the brand, so they required Mondelez to take down the videos and encouraged the company to ensure future influencer collaborations are more clearly labeled as ads.

There are certainly other legal considerations that brands should keep in mind before working with influencers, particularly if the influencers are based in foreign countries. For example, in the UAE, influencers technically require a trade license to comply with the UAE Commercial Transactions Law, even though there has been little enforcement by regulators to date. The law states that any activity resulting in a profit is rendered a business transaction, which follows in legal repercussions and liability subject to the applicable provisions of the relevant laws. Diana Hamade, a local UAE attorney, explains, "The fact that the influencer is convincing the consumer to buy a product, the business activity is rendered an advertisement and generally, advertisements must comply with the relevant advertising standards and consumer protection legislation in each jurisdiction in which they are published/posted."[11]

When working with influencers at home or in international markets, brands should do everything possible to get up to speed on the latest regulations related to influencer marketing. For now, the best guidance for marketers is: when in doubt, #disclose, #disclose, and #disclose again; making all efforts to be transparent about the brand–influencer collaboration should be sufficient in most cases.

What About the Competition? Tackling Competitive Risks

Beyond legal and brand risks, marketers should also be cautious of potential competitive risks that could arise from either an influencer's previous or future brand collaborations (Fig. 10.2). Matt Britton, chief executive officer of the influencer marketing firm Crowdtap, warns brands, "The fact is they [the influencers] can work with your brand today and a competing brand tomorrow – there's nothing to stop them, which diminishes the endorsement."

What this really boils down to is—brands need to make competitive insights part of their criteria for influencer identification, as well as have the right contractual protections in place before they start working with an influencer to ensure the influencer won't sign on with its biggest competitor one week later.

Mae Karwowski, chief executive officer of the influencer marketing technology company, obviously shares the experience of a prominent juice brand client that was concerned about collaborating with wellness-focused influencers who previously worked with other juice brands. "We had to push back and explain to them that if you're a wellness influencer, a lot of content you produce is going to be about juice," she says, going on to describe her firm's process for vetting influencers for competitive issues: "We do scan all

"Brands should be concerned about the past affiliations of the influencer and future unanticipated affiliations."

-Matthew B. Britton (@MattyB)
Chief Executive Officer
CROWDTAP

DigitalInfluenceBook.com   SHARE THIS

Fig. 10.2 Matthew Britton

influencers' photos and videos to see what competitors they're talking about, and how frequently they're talking about them, and then we'll also write competitive issues into contracts with influencers when it's really necessary, and when the brand deems it necessary to have 'exclusivity windows.'"

Outside of her firm's work with the juice brand, Karwowski describes how they address competitive risks more broadly: "We work with Google, so we're going to make sure an influencer isn't talking about Apple. In some industries, competitive concerns are more important than others. For example, the juice company needs to realize that there are other juices being mentioned, unless they want to sign a long-term contract with a specific influencer."

She observes that long-term contracts between brands and influencers are becoming more common. "We have long-term contracts with brands and we want to test and find the influencers who are really performing the best for these brands and then sign the best influencers on for long-term engagements as well – it's in the best interest for the influencer, they're more excited to work with one brand they like for a long time instead of working with all their competitors."

Beca Alexander, president and founder of the influencer marketing agency Socialyte, has this to add: "We have brands come to us with competitive concerns. For example, automotive brands are less likely to partner with influencers who have collaborated with other auto brands – typically influencers can't mention another auto competitor prior to collaboration for at least three months, that way the competitive content is buried in their feed."

Alexander also makes the point that other industries have much narrower competitive concerns: "A spirits brand will only care about other spirits in that specific category. For example, a vodka brand will only want exclusivity from other vodka brands, and will not care about previous influencer collaborations with other spirits such as wine or champagne brands."

Remember from Chapter 6, there are situations when it may be beneficial for an influencer to have experience with many companies in the same industry. Tamara McCleary, a B2B influencer among a chief marketing officer audience, says, "If I only work with a single brand as an influencer in the B2B space, no one is going to listen to what I say. They know it's ridiculous that one brand would have all the answers. The most influential influencers in the B2B space work with all of the companies in a given category as an unbiased source to talk about the solutions and the problems so that their audience will trust them."

Faking Influence: How Some Influencers Deceive Brands

How easy is it for individuals to "buy influence" by paying to gain new followers and boost content engagement rates? In 2017, Mediakix, an American influencer marketing agency, ran a social experiment to find out the answer—specifically, how much money would it cost to create a completely fake Instagram influencer that a brand would be willing to work with. Mediakix created two fictitious Instagram influencer accounts that they grew fully through purchased followers and engagement (such as likes and comments). With an established social footprint, the company then used the fake influencer Instagram accounts to apply for open campaigns that had been published to influencer marketplace platforms.

Here is a sampling of photos posted to one of their fake accounts (Fig. 10.3).

Mediakix hired a local model for a photo shoot and developed all of the Instagram feed content in a single day of shooting. They then gradually posted the images each day to their fictitious Instagram account. At the same time, they purchased anywhere between 1000 and 15,000 followers a day to bulk up their following at a price tag of $3–8 per every 1000 followers.

Once the account hit a few thousand followers, Mediakix also started paying for engagement in the form of likes and comments—for each photo they purchased 500–2500 likes and 10–50 comments. From there, they signed up for a variety of influencer marketplaces, where they applied for open brand collaborations. The "CaliBeachGirl310" account secured a brand deal with a swimsuit company and a national food and beverage company through the platform application process. They did all this for under $1000 and secured their first brand deals worth as much as $500.[12] They made this happen with a completely fictitious account and purchased followers and engagement—it was way too easy for them to game the system. Mediakix revealed their entire step-by-step process to the world, in a blog post it published titled, "How Anyone Can Get Paid to be an Instagram Influencers with $300 (or Less) Overnight."

Evan Asano, chief executive officer of Mediakix explains, "We assumed it was going to be reasonably easy to build these fake accounts, but what really shocked us was how easily we were able to source paid brand deals. We realized we could have easily continued the ruse and continue to grow the accounts and source larger and larger brand deals."

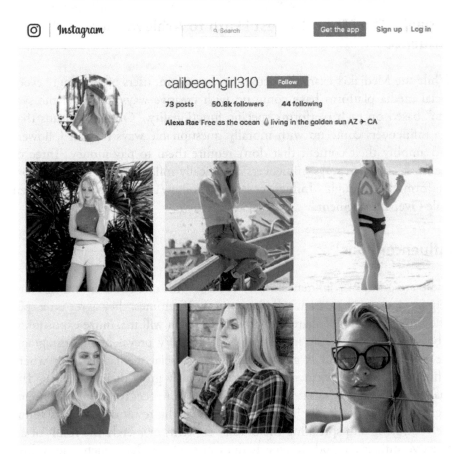

Fig. 10.3 Assorted Instagram photos from Mediakix's fictitious influencer account

Chapter 9 explains the ways brands and influencers can collaborate which is only limited to their creativity, and the same can be said for the many creative ways that some influencers use deceptive tactics to cheat their way to brand collaboration. As the tactics employed by these unscrupulous "influencers" will surely continue to evolve, the following is a sampling of the different tactics that are common in the marketplace at the time of writing this book. Broadly speaking, they fit into the following two categories of "Organic" and "Paid." There are certainly others out there, but these tactics serve as a foundation to understand how some influencers go about gaming the system in their own self-interests.

Organic: The More "Honest" Path to Achieve Fake Influence

While the Mediakix example focuses on Instagram, users of seemingly every social media platform have come up with creative ways to "game the system" based on each platform's specific functionality. "Organic" means that the influencers come up with morally questionable ways to gain followers and amplify their content that don't require them to pay money. Three of the most common ways influencers organically inflate their influence across platforms are through: "Influencer Pods," "Follow/Un-Follow," and "Golden Rule Over-Engagement."

Influencer Pods[13]

These are groups of influencers who collude with one another and agree to help engage and share each other's content. Oftentimes, they agree on a specific time of day to all share the same post which will maximize exposure to a broad audience. While influencer pods are very prevalent on Instagram, the approach can and is applied across any social media platform where influencers want to partner with peers to jointly grow their influence. The challenge for brands is that when they pay an influencer to promote content, the brand does not know if the engagement is the result of followers genuinely enjoying and spreading the brand's content, or merely the result of an effective influencer pod. Maybe it does not matter that much—according to YouTuber and editor Zach Bussey, "If [a brand] had hoped to get [its] message in front of X number of people, if you subtract other pod members, that number is dramatically reduced. That said, it's not impossible for those pod members to be interested in the message – it's just not their reason for being there."[14]

Follow/Un-follow

This tactic is exactly as the name suggests—"influencers" on social media platforms like Twitter or Instagram follow other people they do not know or care about solely because they hope the person will follow back in return. After a few days or weeks, the "influencer" un-follows the person, so their follower count shows a net gain. Matt Navarra, director of social media at The Next Web, sums up how the economics work over time: "If they follow 50,000 accounts in one day, and only 1% follow them back, that's still 500 new followers generated for their account with little, or no effort. Repeat this daily for months or even years, and you can see how they can quickly create an account with millions of followers."[15]

The numbers start getting really big when "influencers" pay for automated follow/un-followers services or even pay freelancers in third-world countries to do it manually on their behalf. Brands can use a variety of social media monitoring tools that are available to analyze an influencer's follower acquisition patterns over time to spot if an influencer is faking it though follow/un-follow.

Golden Rule Over-Engagement

Everyone knows the Golden Rule: "Do unto others as you would have them do unto you." In everyday life, this might mean a friend buys you a birthday gift and then you feel obligated to buy them a gift on their birthday as well. The social media world has its own unspoken Golden Rule—one that is generally based on a combination of "giving first" and "reciprocity." Mark Schaefer, executive director of Schaefer Marketing Solutions, explains, "Whatever power structure exists on the Social Web is often built on a foundation of subtle indebtedness, an ability to create influence through an economy of favors." What this means is if Person X engages with Person Y's social media content, then it is highly likely that Person Y will engage with Person X's content in the future.

The "influencers" sometimes take advantage of the Golden Rule by expending tremendous efforts interacting with and promoting other people's content with two goals: to get on those individuals' radars and convert them to followers, and convert the followers of their new followers into their own followers. Brands can discover these individuals pretty easily by simply reviewing the influencer's recent interactions—if it looks superficial (ex: "Great post" "Nice!"), then they are likely gaining a following that does not genuinely care about what they produce.

Paid: More Money, More "Influence"

As the Mediakix example demonstrates, it is pretty easy for individuals looking for shortcuts to gain their following and increase engagement by buying their way to "influencer" levels. Who can blame them? There are so many paid services and online tools available that help people cheat their way to the top—a simple Google search for "companies to buy social media followers" delivers pages and pages of results featuring companies that help deliver followers and engagement for a fee.

The paid services are very specific, based on the type of unauthentic engagement the individual wants to purchase. An in-depth investigation conducted by the *New York Times* found that one such "pay to play" follower acquisition company called Devumi possesses 3.5 million automated Twitter accounts, each sold multiple times to customers resulting in more than 200 million fake Twitter followers.[16] These services are relatively inexpensive as well—one company called "Buy Real Marketing" promises its customers that they can purchase 500 Instagram followers for $7; 1000 Twitter followers for $12; 10,000 YouTube views for $47; or 2000 Facebook likes for $69. This is not a lot of money to spend if their customer is able to replicate Mediakix's experience and convert the purchased followers and engagement into paid brand collaboration deals brokered by faceless influencer–brand broker online marketplaces. The services are pretty comprehensive as well— the following chart summarizes just some of the many social interactions customers can pay Buy Real Marketing to perform on Facebook, Twitter, YouTube, and Instagram (Fig. 10.4).

These services are generally powered behind the scenes by low-cost workers in developing countries or "bots" to drive the desired online behavior. For example, throughout Asia, there are a series of operations commonly referred to as "click farms," where local workers operate hundreds of cell

Facebook	Twitter	YouTube	Instagram
Post Likes	Followers	Regular Views	Followers
Shares	Retweets	Comments	Likes
Comments	Likes	Subscribers	Comments
Followers	Daily Followers	Likes	Daily Likes
Five Star Ratings	USA Followers	Favorites	Daily Followers
"Post Haha Reactions"	Account Management	Shares	Video Views
*From BuyRealMarketing.com website			

DigitalInfluenceBook.com   SHARE THIS

Fig. 10.4 Fake influencer service offering

phones and manually perform requested tasks on social media such as gaining followers or driving a certain type of engagement. In June 2017, Thai police raided a click farm operated by three Chinese nationals in a rented home near the Cambodian border. In total, Thai police seized 474 iPhones and 347,200 SIM cards from Thai mobile phone operators.[17] The click farm workers told police that a Chinese company supplied all of the smartphones and paid them each 150,000 baht (approximately $4403 USD) per month. They chose Thailand as their base of operations because of the country's relatively inexpensive mobile usage fees.

In addition to artificially inflating their own online engagement, when "influencers" use these paid services, there can also be unexpected negative consequences that could hurt any brand who hires them. Jim Harris, a Canadian management consultant and leading B2B influencer for major global brands on topics like "disruptive innovation," recently attended an event and noticed that one of the other participants was outperforming him in terms of reach and engagement on Twitter. When Harris approached the participant to see what was leading to his high-performance levels, the participant told Harris that he had a bot specifically designed to retweet any Twitter post with the event hashtag. This is a common mistake Harris sees occur at major industry conferences: "Upon a closer examination of his Twitter feed, it was clear that his bot had inadvertently retweeted several "hashtag hijackers" such as escort services and other unsavory businesses who used the hashtag to catch the attention of event attendees and get their business." The conference organizers probably did not appreciate the unwanted brand association with these other businesses all across the Twitterverse.

Be Aware of the Dark Side and Protect Your Brand

Whether its brand damage that could be caused by working with the wrong influencer (ex: Disney-PewDiePie) or missed opportunity caused by paying a so-called influencer for their fake influence (ex: Mediakix's fake Instagram account experiment), there are many potential risks and pitfalls that brands may encounter when they implement their influencer strategy. From a business perspective, working with influencers can expose the brand to unintended reputational, legal, or competitive risks that result from the influencer's actions. Meanwhile, brands can easily blow through their influencer marketing budgets by working with "influencers" who have relied on buying influence through morally questionable practices—either organic or paid. It is up to key brand personnel on the marketing, public relations, or other functional team responsible for influencer activities to be aware of the "dark side" and take necessary precautions to verify the influence of their brand partners and monitor partnering influencers for future activities that could negatively impact the brand.

What does all of this "fake influence" mean for brands that are trying to measure the effectiveness of their influencer marketing programs? Turn to the next chapter, *Measure Success—What's the Return on Investment* to find out.

Marketer's Cheat Sheet

- **The Two Dark Sides**: The two dark sides of influencer marketing refer to one set of potential pitfalls related to the brand's risk exposure (such as reputational, legal, and competitive risks) and one set related to the many ways supposed influencers can trick brands into working with them through faking influence (either organically through measures like pods and follow/un-follow or paying for follower acquisition and account engagement services).
- **The Dark Side is Constantly Evolving**: Unscrupulous influencers are constantly on the lookout for new methods to exploitatively gain their following and boost engagement on their social media platforms of choice. At the same time, those social media platforms regularly work to identify loopholes and adjust algorithms to penalize "fakers" and promote the truly influential. This means brands need to stay up to date on the latest tricks of the trade and monitor influencer performance carefully.
- **Legal Frameworks Are Playing Catch-Up**: As governmental agencies and consumer advocacy organizations learn more about influencer marketing, the rules and regulations that brands need to abide by in different countries continue to evolve. For now, brands in most countries like in the USA, UK, and across the EU simply need to make sure their influencers use the appropriate disclosures whenever possible to ensure their collaboration cannot be perceived by outsiders as deceiving local consumers.
- **Competitors Can be Good and Bad**: When it comes to influencer marketing, competitive dynamics can be more complex. While no brand wants to see itself mentioned by an influencer who is also endorsing an archrival brand, marketers should keep an open mind in certain cases where the influencer working with multiple brands lends to their objectivity/credibility among their audience, such as the juice company or Tamara McCleary's B2B technology brands.
- **Don't Take Numbers at Face Value**: When identifying influencers to partner with, never take their number of followers or engagement rates at face value. There are so many tools out there that help brands authenticate whether an influencer is gaming the system—remain vigilant and avoid paying hundreds or thousands of dollars for fake social media accounts to promote content to fake followers.

Notes

1. Barnes, Brooks. "Disney Buys Maker Studios, Video Supplier for YouTube." *The New York Times*, 24 Mar. 2014, www.nytimes.com/2014/03/25/business/media/disney-buys-maker-studios-video-supplier-for-youtube.html.

2. Google. "Multi-channel Network (MCN) Overview for YouTube Creators—YouTube Help." *Google*, support.google.com/youtube/answer/2737059?hl=en.

3. Barnes, Brooks. "Disney Buys Maker Studios, Video Supplier for YouTube." *The New York Times*, 24 Mar. 2014, www.nytimes.com/2014/03/25/business/media/disney-buys-maker-studios-video-supplier-for-youtube.html.

4. Winkler, Rolfe, et al. "Disney Severs Ties with YouTube Star PewDiePie After Anti-semitic Posts." *The Wall Street Journal*, Dow Jones & Company, 14 Feb. 2017, www.wsj.com/articles/disney-severs-ties-with-youtube-star-pewdiepie-after-anti-semitic-posts-1487034533.

5. Roettgers, Janko. "Disney's Maker Studios Drops PewDiePie Because of Anti-semitic Videos." *Variety*, 13 Feb. 2017, variety.com/2017/digital/news/disney-pewdiepie-anti-semitic-videos-1201987380/.

6. Patel, Sahil. "Inside Disney's Troubled $675 Mil. Maker Studios Acquisition." *Digiday*, 22 Feb. 2017, digiday.com/media/disney-maker-studios/.

7. Roderick, Leonie. "L'Oréal Faces Backlash After Dropping Influencer Following 'Racist' Comments." *Marketing Week*, 1 Sept. 2017, www.marketingweek.com/2017/09/01/loreal-faces-backlash-diversity/.

8. Norton, Quinn. "Robert Scoble and Me—Quinn Norton—Medium." *Medium*, 19 Oct. 2017, medium.com/@quinnnorton/robert-scoble-and-me-9b14ee92fffb.

9. Poggi, Jeanine. "Google, Awesomeness TV and Maker Studios Come Under Fire for Influencer Marketing to Kids." *AdAge*, 21 Oct. 2016, adage.com/article/media/google-awesomeness-tv-maker-studios-fire-influencer-marketing/306405/.

10. ASA.org.uk. "ASA Adjudication on Mondelez UK Ltd." *ASA.org.uk*, 26 Nov. 2014, www.asa.org.uk/rulings/mondelez-uk-ltd-a14-275018.html#.VHWXYNKsWN0.

11. Hamade, Diana. "UAE Law and Influencers: What You Need to Know." *ITP Live Middle East*, 6 Apr. 2017, www.itpliveme.com/content/394-legal-guidelines-you-need-to-know-about-influencer.

12. Mediakix Team. "How to Be an Instagram Influencer For $300: A 2-Month Study." *Mediakix*, 4 Aug. 2017, mediakix.com/2017/08/fake-instagram-influencers-followers-bots-study/#gs.jW1X3Fw.

13 While Influencer Pods are the result of organic influencer collaboration, many "exclusive" pods exist where the organizing influencer requires participants pay a fee to join the pod.

14. Pathak, Shareen. "Podghazi: Instagram Influencers Use Comment Collusion to Game the Algorithm." *Digiday*, 26 July 2017, digiday.com/marketing/podghazi-instagram-influencers-use-comment-collusion-game-algorithm/.
15. Navarra, Matt. "Twitter Is Being Spoiled by One Type of User." *The Next Web*, 18 Aug. 2016, thenextweb.com/twitter/2016/08/18/twitter-follow-unfollow-spam-sucks/.
16. Confessore, Nicholas, et al. "The Follower Factory." *NYTimes.com, The New York Times*, 27 Jan. 2018, www.nytimes.com/interactive/2018/01/27/technology/social-media-bots.html.
17. Deahl, Dani. "Three Men in Thailand Reportedly Ran a Clickfarm with Over 300,000 SIM Cards and 400 iPhones." *The Verge*, 12 June 2017, www.theverge.com/2017/6/12/15786402/thai-clickfarm-bust-iphones.

11

Measure Success: What's the Return on Investment?

Kimberly-Clark is one of the world's largest consumer packaged goods companies, specializing in personal care products by brands like Kleenex, Cottonelle, Huggies, Scott, and Kotex. Jason Davis, a 20-year marketing veteran, works for them as "Senior Brand Manager, Shopper Marketing," and while he does focus on many aspects of his employer's consumer-facing marketing efforts, these days he finds himself spending more and more time on advancing its influencer marketing initiatives. The importance of influencer marketing for Kimberly-Clark has grown exponentially in recent years, as its consumers increasingly turn to trusted online Category Influencers to help inform their purchase decisions—this is especially true for many of their products, since items like Depends adult diapers or Kotex tampons are not typically discussed during dinner parties in most consumers' social circles.

Despite the clear link between its influencer marketing efforts and the brand's ability to engage target consumers, Davis still struggles to make the case for investment in future influencer marketing activities. In an interview with industry publication *eMarketer*,[1] he says, "If we're not generating sales, we're not going to have the funds to put toward influencer marketing."

As with many executives charged with building influencer programs from the ground up, Davis needs to educate internal stakeholders across departments—particularly senior management—about the value influencer marketing generates, as opposed to allocating more marketing budget to traditional advertising channels like magazines and television advertisements.

© The Author(s) 2018
J. Backaler, *Digital Influence*,
https://doi.org/10.1007/978-3-319-78396-3_11

Davis explains: "We rely on vanity metrics, but we're pushing hard to get beyond those and tie influencer marketing to sales." Vanity metrics are easy to report on and include standard social media metrics such as impressions, likes, or shares, but rarely do these high-level metrics allow marketers like Davis to report back to management how his spending on influencer marketing ties back to sales of Kimberly-Clark's products.

What regular vanity metric reporting does achieve is help to provide management (often times much older and lacking "digital savvy") with an easily understandable benchmark to understand "How are we doing? Better or worse?" But, most industry experts agree that reporting on vanity metrics alone will not satisfy senior management for much longer—especially as "heavy spender" industries like fashion, beauty, toys, consumer electronics, alcohol, fitness, and wellness spend millions on influencer marketing each year.

Through creative use of influencer-specific coupons, trackable links, and promotional codes, Davis is moving to shift the Return On Investment (ROI) conversation away from vanity metrics and toward sales. He gives the following example: "An influencer shares a promo code, and when consumers use that code on Kimberly-Clark products, we can track it back to a specific influencer activation. Then we can say that code ABC drove X many sales. Otherwise, with all the activity going on in the retail universe, it's hard to attribute sales to a specific influencer activity versus another sales driver." Being able to tie an influencer's activities directly to sales is the most straightforward way to justify ongoing spending on influencer marketing to management.

It is not always easy to accomplish, but Kimberly-Clark has developed a series of creative ways to attribute influencer collaboration to product sales. When it came time to promote its "U" by Kotex specialty fitness tampon brand, the company engaged Category Influencers like Courtney Danielle, a blogger and YouTube creator at "Curls and Couture," who focuses on beauty and women's health. Davis explains, "In this area, convenience and discretion are important. Consumers want to talk to people who are going through the same issues that they are."

When Danielle's audience views her YouTube product promotion video,[2] a pop-up for Kimberly-Clark's product appears directly below the video, encouraging viewers to purchase online from the American retailer Target. The viewer is redirected to the product page on Target's online store through a specialized trackable link.

As a major retail channel of Kimberly-Clark products, Target can share purchase data with the Kimberly-Clark marketing team to help tie its influencer spending to final product sales. This closed-loop approach helps the Kimberly-Clark marketing team understand how Danielle's (or other influencer's) videos contribute to awareness, clicks, and ultimately new product sales.

In another instance, Kimberly-Clark partnered with the influencer marketing agency Linquia[3] to boost sales of its Cottonelle brand. The two firms worked together to launch an influencer shopper program that featured a deluxe-size sample box of a wide assortment of personal care items like tissues, lip balm, and hand cream in a special box called the "Cottonelle Clean Care Box." Linquia paired Cottonelle with more than 30 parenting-focused Category Influencers who demonstrated an interest in home and body care products. The influencers similarly drove their audiences to Target.com, where consumers could purchase their own Cottonelle Clean Care Box for a limited time.

While the Kimberly-Clark team could boast vanity metrics that the campaign achieved 23,000 "online engagements" or an "audience reach" of six million, the number one thing that mattered for their continued investment in influencer marketing was the fact that within 48 hours of the promotion kickoff, all Cottonelle Clean Care Boxes sold out on Target.com.

As Davis advances the role influencers play in Kimberly-Clark's consumer marketing strategy, he will need to continue to tie as much activity as possible to sales, as was the case in the two preceding examples. However, ROI is not always about sales; it ties back to the original goals that a company establishes at the outset of their influencer marketing program. While sales are the basis of the program for Kimberly-Clark, others may care more about different outcomes. This chapter helps shed light on how companies can bring more measurement to what is often perceived as an unmeasurable art.

Common Ways Brands Measure the ROI of Influencer Marketing

Ultimately, determining the ROI of influencer marketing efforts ties back to whatever the brand's original goals were when they first started the program. A one-off influencer campaign to launch a new product will have different criteria to measure ROI than a long-term oriented B2B influencer relations program designed to improve corporate reputation. Carlisle Campbell, senior director of communications at Capital One, sums it up quite nicely: "First, think about the actual objective you are trying to achieve. What do you want influencers to do? Then, set KPIs you can achieve, and find what's working and what's not.[4]"

Chapter 9 introduces many of the initiatives brands collaborate with influencers on, including corporate reputation management, amplification of new product launches and brand events, product development inspiration, product reviews, and search engine optimization. Depending on which of these areas a brand chooses to partner on with an influencer, the metrics they report on will vary substantially.

Here is an example metric for each of the activities just listed:

- **Corporate Reputation**
  Example Metric: # of positive media mentions associated with influencer engagement
- **New Product Launch**
  Example Metric: total sales generated by influencer-specific promotion code
- **Product Development**
  Example Metric: # of new ideas generated from influencer focus group sessions
- **Product Reviews**
  Example Metric: # of product reviews by known influencers on Product X product page
- **Search Engine Optimization**
  Example Metric: improvement in keyword ranking from influencer-produced content

According to the American Marketing Association, "It's important to understand that marketers don't use influencer marketing solely to drive product sales. Influencer marketing can impact everything from top-of-the-funnel metrics like awareness and brand perception to mid-funnel metrics like email newsletter registrations, coupon downloads and contest entries. While influencer marketing can also be very effective at driving product sales, that's not always what marketers are using it to achieve.[5]"

Vanity Metrics: How Important Are They?

Where brands tend to fall short is when they do not set the right goals at the outset, or blindly decide to "test out influencer marketing" because it seems like the new hot thing to try. "I've seen situations where after influencer campaigns, brands come back and say, 'We've got a lot of numbers, but we didn't see where it had any impact on our business,'" explains Chris Gee of the global public relations firm Finsbury. "That's usually where they weren't looking at the right criteria in selecting influencers, they weren't looking at deploying the influencer in the right way, and in a lot of cases – they simply don't understand their audience."

Oftentimes, the numbers a brand receives from their team or outside agency in the situation Gee describes are what practitioners call "vanity metrics" (also known as "engagement metrics" or "consumption metrics"). Here are a few examples:

- Number of likes
- Number of shares
- Number of comments
- Number of followers
- Number of downloads
- Number of page views

These metrics are relatively easy to report on. Brands and influencers can share their social media dashboards to easily convey most common vanity metrics requested. But the question marketers need to constantly ask themselves is—are vanity metrics the *right* metrics, or are they simply the easiest ones to report on?

In a blog post, the prolific marketer Seth Godin[6] explains the difference:

> When you measure the wrong thing, you get the wrong thing. Perhaps you can be precise in your measurement, but *precision is not significance.*
>
> On the other hand, when you are able to expose your work and your process to the right thing, to the metric that actually matters, good things happen.
>
> We need to spend more time figuring out what to keep track of, and less time actually obsessing over the numbers that we are already measuring.

There are times, however, when vanity metrics may be the right choice, or at least the best option available, to measure the ROI of a brand's influencer marketing efforts. For Jason Davis at Kimberly-Clark, ongoing reporting on vanity metrics allows him to provide management with a framework to assess his team's performance. While Davis ideally wants to tie more of his team's influencer marketing activities directly to product sales, for now he can use vanity metrics in to achieve management alignment and get additional budget to continue to test and expand.

Alternatively, some brands are perfectly satisfied reporting on vanity metrics alone. "A lot of clients come to us and say, 'We want you to help us build buzz around our brand,' so they hire us to engage influencers on their behalf to 'build buzz,' which is generally measured in the forms of likes, shares, and general online engagement," shares Brendan Gahan, founder of EpicSignal, a digital video-focused agency based in New York.

38%

"38 per cent of marketers say they are unable to tell whether influencer activity actually drives sales," according to Rakuten Marketing's study of 200 influence marketing practitioners across the U.K.

As influencer marketing becomes a more strategic aspect of brands' marketing mixes, vanity metrics alone will become less and less significant, or as Daniel Hochuli, content marketing manager at LinkedIn, explains, "For business goals, such as ROI, vanity metrics should take a back seat to those metrics that build the customer lifetime value narrative (conversions, subscriptions, MQLs, SQLs, etc.). But note, this is not a quick win. CLTV takes time, A/B testing, volumes of content, and conversions to build an accurate picture. Don't look for the quick and dirty win with vanity metrics; it's not there.[7]"

Generational Conflict: Operators vs. Budget-Holders

Don't forget this key takeaway from Chapter 2: While word of mouth has been around since early man gave the first recommendation to a trusted peer, what has changed significantly is the role word of mouth plays in a world where individuals are connected to more like-minded peers than ever before, and internet-enabled word of mouth spreads messages and builds influence at never-before-experienced speed.

It is the "internet-enabled" component which makes modern-day influencer marketing much more challenging for many brands to implement—especially in long-established multi-national companies where senior leaders who make budget decisions may not possess the digital savviness necessary to realize the potential of how influencers producing content on a diverse range of unfamiliar online platforms could add value to marketing performance. This is in contrast to taking the same budget and investing in traditional areas such as television ads, print placements, or potentially online banner ads on the digital version of traditional print media outlets.

This is why Tim Williams, chief executive officer at the influencer marketing software firm Onalytica, believes education for senior management is really key: "What's happened is the amount of social and digital channels has really increased over the past 10 years or so. As a result, a lot of senior management don't really understand what's possible."

Williams believes that when it comes to determining ROI it is just as much an upper management issue as it is a measurement issue: "Most people who are digital or social experts know how influencers are behaving on these channels, but they're struggling to put together influencer marketing-programs, because they're not being measured on this and senior managers don't really understand how to manage outcomes from this, because they don't understand it as much. It's one of the cultural reasons why some influencer programs don't get off the ground."

James Hare, vice president of marketing[8] at a Fortune 500 multi-national in the B2B space, leads influencer relations for the company and validates Williams' view: "One of the challenges with influencer marketing is our most-senior executives didn't grow up in a digital world with social media. Their understanding tends to end at things like impressions – what I call vanity metrics."

Hare does not really care about vanity metrics, but he reports on them because that is what his bosses understand. "I care about those metrics because that's how I get budget (I say, 'We're going to go to this event, and we're going to generate this many impressions'), but what I really care about is how the people who see our influencer-generated social media postings at the event are going to click-through to our content? As they click through – how many people are going to register to download our content, and how many registrations are going to turn into validated leads?"

Hare believes it is extremely important for anybody who is setting up an influencer marketing program to have a combination of metrics—one set that is easily understood at the management level and another set that really matters for moving the business forward. "When it comes to getting budget, it's understanding which metric is going to be most essential to the people making a decision on the budget," Hare says. "I think where people run into problems (those who say influencer marketing didn't work for us) is because their executive team said, 'We need to see this lead to X number of opportunities in year one.' That's not how this works – this is a time play, something that needs six months to a year – actual timing is different for different businesses – until you're going to see the results come in. So, knowing what those metrics are and then putting a process in place and the right expectations on timeline for achieving those metrics. It's a lot of trial and error – you just don't know how long it will take for the results to start coming in."

One way brands are attempting to define influencer marketing impact in "traditional terms" that may be more relatable to senior management is by using Advertising Value Equivalent (AVE). AVE began as a way to measure the impact of PR agencies, and it refers to the cost of buying the space taken up by a piece of media coverage, had that coverage been an advertisement. Christian Damsen, senior vice president of the influencer management software firm Traackr, feels that using advertising metrics like AVE to quantify influencer impact is leading to bad habits in the industry: "I regularly see brands use metrics like AVE, but these are really the wrong types of metrics, because an influencer is not an ad impression."

Damsen is not alone in his skepticism regarding the use of AVE as a measure of influencer impact. The International Association for the Measurement and Evaluations of Communication (AMEC) released a set of principles, commonly referred to as the "Barcelona Principles," that firmly discourage brands from using AVEs, since there is little evidence that suggests an actual link between the two. Marketers should do their best to develop an understandable set of metrics for senior management, but they also need to choose wisely, and not go with something like AVE that does not truly communicate the impact of influencer marketing.

Step-by-Step, ROI Is Becoming Clearer

As marketers gain experience with influencer marketing in general, and how to build consensus and gain necessary budget from their internal management specifically, they are gradually developing more sophisticated ways to measure the impact of their efforts. The last section in this chapter offers a practical framework for measuring ROI, and the following are a series of examples that demonstrate how a few organizations are increasing their ability to tie influencer marketing activities to ROI based on their brand's goals for influencer engagement.

Amber Armstrong, who developed the influencer relations program for IBM Watson Customer Engagement, has witnessed firsthand how ROI measurement is evolving for B2B companies: "We're at an inflection point where companies are transitioning from influencer marketing ROI centered on awareness-focused metrics to actually having to show that it's driving results for the business."

She offers more detail about how her team is making this transition: "While we want to keep the general awareness piece, but we also want to add the demand generation piece. We do this in a few different ways. One way is by working with influencers to host webcasts or author e-books. The influencer creates a webcast, invites their audiences (it's clear that its sponsored by IBM), we bring in select IBMers to share some of our related content, and then IBM lead development reps place follow up calls to the attendees. We drop the audiences into nurture streams with email content. Moving from a very high-level awareness driven, to really being much more analytical and results-based." Through this systematic approach, IBM Watson Customer Engagement is able to link influencer activities directly to lead generation and ultimately, to sales (Fig. 11.1).

Mae Karwowski, whose influencer marketingtechnology company Obviously works with many top consumer brands including Coca-Cola, Sephora, Heineken and Uniqlo, says that a lot of the influencer client work her firm focuses on is top of the funnel brand awareness. While it can be challenging to assign metrics to measure awareness-related goals, her firm has developed creative solutions to help their clients gain a better sense of how their influencer activities are performing. When working with influencers whose primary platform is Instagram, Karwowski and her team calculate media value by using Instagram benchmark CPMs for the related industry. "We've found clients receive three times their ad spend compared to reaching the same number of people using Instagram's paid advertising services," she says.

This approach differs from the generic AVE approach (see previous section), because brands can clearly benchmark influencer performance against the cost of paid ads on the same platform: "This is becoming very common and brands are very receptive to this approach. We also drive lead generation

"Influencer marketing is becoming a much more sophisticated discipline within the marketing department. We typically find that IM content generates 2-4 times ROI and more importantly, drives even higher engagement."

-Amisha Gandhi (@amishagandhi)
Head of Global Influencer Marketing
SAP

DigitalInfluenceBook.com   SHARE THIS

Fig. 11.1 Amisha Gandhi

through email sign-up campaigns and track revenue through unique URLs – then we can look at cohorts of emails to see which influencers are performing the best in terms of revenue." This approach is extremely effective for tracking ROI for consumer-facing brands that sell direct to consumer via e-commerce. However, for brands with a long history of selling through traditional retailers like Kimberly-Clark, they need to come up with more complex tracking systems to tie influencer activities to sales.

It is not just the approach to ROI measurement that is advancing, but also the expectations brands have for the impact they generate through influencer collaboration. Dez Blanchfield, a top technology and big data influencer, shares how the brands he works with are demanding more from their influencers: "A lot of brands invite influencers to attend their conference and events hoping that they will amplify key takeaways on social media. The reality is, many of these influencers will show up all-expenses-paid, send off a few tweets and then call it a day."

He explains that more experienced brands are starting to focus much more on quality of engagement, rather than simply being satisfied with just having the right big-name influencers in the room: "When a company pays for me to attend their event, I consider it my responsibility to work the whole time I'm there. I live-cover the entire event, stick around late to interview speakers when they're finished, and even provide the brand executives with daily metrics reports to help them communicate the ROI the inclusion of influencers at the event generated to their management team."

Examples like these are helpful to see how the ROI conversation is shifting from simple reporting on baseline vanity metrics to more tangible business outcomes like the number of leads generated or total products sold. This shift is easier for companies to make that have already invested in influencer marketing—they have most likely failed a few times along the way, and if they are foreword-thinking have applied those learnings to increasingly sophisticated measurement approaches—but what about brands that are getting started for the first time? How can they do their best to align influencer marketing measurement to business goals from the outset and avoid some of the costly mistakes encountered by early experimenters? There is a framework for that—it is all about measuring inputs, outputs, and outcomes.

Measurement Framework: Inputs, Outputs, Outcomes

The International AMEC, introduced earlier in this chapter, developed an Integrated Evaluation Framework[9] to provide a consistent and credible approach to help organizations bring measurement to business areas like public relations that are often difficult to quantify impact. Their approach focuses on linking clear objectives to a series of related inputs, activities, outcomes, and outputs to achieve desired impact. Although the AMEC model is complex, the marketing software firm Traackr[10] does a good job adapting the original framework to meet the needs of today's influencer marketing practitioners.

Their approach begins by defining a common set of measurement terms (Fig. 11.2).

The three components of their measurement framework are inputs (program activities), outputs (proxy to business outcomes), and outcomes (impact on business outcomes). Specifically, input metrics help marketers measure the quantifiable activities or touch points across the brand's influencer marketing program. Output metrics are marketer's proxy to their brands marketing and business outcomes. Lastly, outcome metrics are directly derived from a marketer's marketing or business goals—again simpler than the AMEC model, but still a little difficult to understand at face value. That's why the following illustrative example makes this a whole lot clearer.

The following "Sample Influencer Marketing Measurement Matrix" breaks down specific types of metrics that brands can use to measure input metrics, output metrics, and outcome metrics that all tie to specific influencer program goals (Fig 11.3).

The matrix starts with influencer program goals that can include managing brand reputation, expanding brand awareness, increasing brand

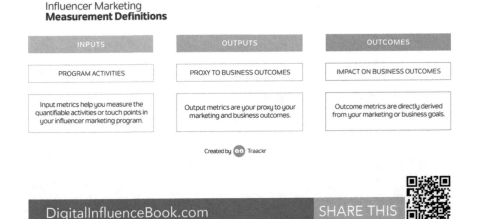

Fig. 11.2 Traackr ROI framework 1

advocacy, driving lead generation, or improving sales conversion (of course, there are many other potential goals depending on what the brand wants to achieve). From there, metrics are assigned across each of the three areas of inputs, outputs, and outcomes, which help serve as a basis for measurement of the brand's goals for its influencer marketing program. Taking a closer look, if a brand wishes to focus on attributing influencer activities to improvements in sales outcomes, then potential metrics include:

Input Metrics:

- Number of influencers engaged online
- Number of influencer guest posts on brand's blog
- Number of influencer product sends

Fig. 11.3   Traackr ROI framework 2

Output Metrics:

- Influencer articles about the brand/product
- Influencer brand/product mentions
- Linkbacks to brand assets by influencers

Outcome Metrics:

- Unique new visitors to owned property
- Sales opportunities generated/carts started

This framework is simply a tool to consider. It should not be viewed as the only way to measure the impact of brand–influencer engagement, but it is a relatively easy to implement solution, especially as brands begin to develop their influencer marketing programs to establish a baseline for measurement from the outset.

Unlocking Influencer ROI: Part Art, Part Science

Ask any influencer marketing practitioner about what it takes to be successful, and they will likely explain that it is part art and part science. Much of influencer engagement to date has been "unmeasurable" from a science perspective, as brands tend to rationalize the impact influencers have on their business by vanity or general "soft metrics." And while there is a place for vanity metrics in some cases (e.g., if a brand truly does want focus on "buzz building"), the vast majority of marketers would be better suited taking a more quantitative approach that ties influencer performance to a thoughtfully designed set of goals.

Some companies are already increasing their sophistication and incorporating tactical elements like coupon codes, trackable links, walled content, and other digital best practices to link influencer activities to results. However, there are too many other companies (especially outside of influencer-intensive consumer industries like beauty and fashion) that need to adopt a more rigorous approach to influencer marketing management. The Influencer Marketing Measurement Matrix introduced in this chapter provides a useful framework to consider when companies are first getting started, or in the middle of an influencer marketing metrics "reboot."

The existing options currently available for marketers to measure the impact of their influencer engagement leave much room for improvement.

Complex systems combining codes to links to spreadsheets that feed into online dashboard interfaces are not the ideal way for marketers to measure their results. Looking ahead, Chapter 13: *What's Coming—The Future of Influencer Marketing* introduces some of the ways emerging technologies may make tracking ROI much easier—by both automating a lot of the manual activities done today and providing a better closed-loop process for brands to weed out fake influence and gain the meaningful insights they are really looking for that matter for the goals they set at the outset of their influencer program.

Marketer's Cheat Sheet

- **Start with Goals**: While marketers rush to report vanity metrics to justify influencer marketing spend, they need to stop and assess how the influencer activities tie back to why they started the influencer program to begin with. It all comes down to goals, and those goals will vary from company to company, and even across different stages of the customer journey.
- **More than Vanity**: When it comes to measuring the return on investment of influencer marketing, brands have tended to overly rely on vanity metrics like clicks and impressions. While vanity metrics have their place for broad brand awareness efforts, most companies would be better off tying influencer activities to business outcomes like lead generation or sales (like Kimberly-Clark is doing with trackable links and coupon codes).
- **Generational Complications**: All too often, the individuals charged with building an influencer program are at least one step (if not multiple steps) down from the executive ultimately responsible for budget. The budget holder tends to lack the digital savvy required to fully understand the power of influencer engagement in why they should spend budget on it. That is why marketers should make sure they present a set of metrics that are relatively understandable (like vanity metrics), but also focus on driving the program forward by tying results to less easily understandable metrics (like content downloads and trackable link attribution)—similar to what James Hare did at his B2B firm.
- **Learning While Doing**: As more brands experiment with different approaches to measuring influencer activities, they are adapting measurement models based on learnings of successes and failures. As a result, more companies are developing systems to attribute influencer activities to goals like sales and lead generation. This has certainly been the case for IBM Watson Customer Engagement and their shift to a more demand generation-oriented approach to measuring influencer ROI.
- **Inputs, Outputs, Outcomes**: There are more tools and measurement frameworks becoming available for marketers to more effectively measure their influencer activities. The Influencer Marketing Measurement Matrix produced by Traackr is one example that brands can consider to assign clear metrics across each stage of their influencer engagement programs that tie back to their original goals.

Notes

1. Minsker, Maria. "For Kimberly-Clark, Coupon Codes Are a Step Toward Tying Influencer Marketing to Sales." *EMarketer*, 23 Dec. 2016, www.emarketer.com/Interview/Kimberly-Clark-Coupon-Codes-Step-Toward-Tying-Influencer-Marketing-Sales/6002267.
2. Danielle, Courtney. *My Top 5 Period Hacks ft. U by Kotex, Curls and Couture*, 29 Mar. 2017, https://www.youtube.com/watch?v=Ks40Ai7SMac.
3. Linquia. *Cottonelle*, http://www.linqia.com/customers/cottonelle/.
4. Traackr. "The Quick Guide to Influencer Marketing Measurement." *Traackr*, www.traackr.com/influencer-marketing-measurement.
5. Steimer, Sarah. "Poll: Measuring Influencer ROI Is Top Marketing Challenge in 2017." *American Marketing Association*, 26 Sept. 2017, www.ama.org/publications/eNewsletters/Marketing-News-Weekly/Pages/Poll-determining-influencer-marketing-roi-is-top-2017-marketing-challenge.aspx.
6. Godin, Seth. "Numbers (and the Magic of Measuring the Right Thing)." *Seth's Blog*, 20 Apr. 2016, sethgodin.typepad.com/seths_blog/2016/04/numbers-and-the-magic-of-measuring-the-right-thing.html.
7. Hochuli, Daniel. "Vanity Metrics: Smart Ways to Use Them." *Content Marketing Institute*, 16 Oct. 2017, contentmarketinginstitute.com/2017/10/ways-use-vanity-metrics/.
8. Interviewee Wished to Remain Anonymous.
9. Bagnall, Richard. "Integrated Evaluation Framework by AMEC." *AMEC*, amecorg.com/amecframework/.
10. Traackr. "The Quick Guide to Influencer Marketing Measurement." *Traackr*, www.traackr.com/influencer-marketing-measurement.

12

Case Studies: Influencer Marketing Best Practices from Around the World

When global consumer technology company Lenovo set out to launch its new YOGA line of products, Chief Marketing Officer David Roman knew that influencers would be an integral part of the global launch, helping to build awareness and ultimately drive product sales in key markets. The products—particularly the tablet which was part laptop and included unique features like a built-in screen projector—were unlike any other product on the market at the time. Aligning YOGA with the right local trendsetters would be critical in communicating this differentiation to its target international, millennial consumer audiences when the product launched worldwide.

Quinn O'Brien, Roman's vice president of global brand strategy, content, and design at Lenovo explained to AdAge[1]: "Design trendsetters are the people in the social group who care what others think about them, and they buy products and get close with brands because those brands say something about them. It's as simple as that. They are a large segment, they are a powerful segment in terms of their spend and buying power, and their friends and peers will look at their products and say, 'What is that?' and they will feel like they are leading a trend."

Roman's internal agency team worked closely with ad agency MullenLowe to research potential millennial online trends to use as the basis of the YOGA product marketing program. Through social listening, they identified an Internet hashtag—#Goodweird—that millennials were using to describe something that was strange, but in an interesting way.

© The Author(s) 2018
J. Backaler, *Digital Influence*,
https://doi.org/10.1007/978-3-319-78396-3_12

"The way this hashtag was being used online aligned with the YOGA's market positioning—for someone accustomed to a standard tablet or PC, the product's design and features are a little strange and unexpected, but the result is a more positive user experience," Roman says. "We embraced the hashtag through developing a dedicated micro-site where users could post their own #Goodweird photos and launched an extensive global content marketing effort and influencer engagement campaign all centered on the #Goodweird theme."

Roman goes on to explain how Lenovo approaches global campaigns: "The key thing is you don't want to go for the lowest common denominator. If you start to think of what's not going to work in some countries, then you would end up with material that's not good enough. What we look at is a concept that's broad enough to cover everywhere, and have the implementation of it owned by each of the country teams so they can make sure the types of humor, the imagery that's used, etc. is relevant in each country. What we don't do is find the lowest common denominator that can be applied safely in every country, because that type of content will not be impactful and breakthrough the noise. We make sure that even though the concept is global, the actual implementation is done in country by the country team."

For the influencer component of the #Goodweird campaign, Lenovo worked with YouTube and influencer marketing agency Portal A, to identify video creators whose audiences were tech-savvy millennials based in key markets where the YOGA products would launch. After an initial round of vetting with YouTube and deep research into each influencers' background and content history, Lenovo's team cultivated relationships with three Category Influencers: EeOneGuy (audience base in Russia and Eastern Europe), The Viral Fever (audience base in India), and Matthew Santoro (audience base in North America). Instead of paying the influencers to be part of the campaign, Lenovo provided them with extensive production resources to develop videos that pushed the limits of their creativity, which was enough to incentivize the three influencers to participate.

To start, each influencer developed their own #Goodweird video specifically for their audience. Lenovo's team provided a framework for production, and ongoing guidance throughout the process, but it was really up to the influencers to come up with their own unique angle that would appeal to their online followings.

EeOneGuy is an eccentric Ukrainian-Russian millennial who produces outrageous comedy videos that help translate trends and current events from the West for an Eastern European audience in a comedic fashion. His video[2] opens with him watching a YouTube video on his Lenovo Yoga, which sucks him into an outrageous techno-pop music video with a series of memorable scenes, including him getting a massage by a room full of overweight Russian men in a Russian bathhouse while sporting a full-body lime green jumpsuit and a group robot dance sequence with him on stage accompanied by female models.

The second component of the influencer video collaboration included all three YouTubers in a dance-off video to engage their respective audiences with one piece of more broadly, globally relevant content. The film opens with the group wrapping up a video conference call, but instead of hanging up they each press a magical #Goodweird button on their Lenovo Yoga device, pitting them against one another in a global dance-off across India, Russia, and the USA, where each is based. Their #Goodweird "Dance Off" video juxtaposes epic Bollywood dance sequences against choreographed American cheerleader routines and frigid outdoor group-dancing in a Russian square. As the film progresses, their cultures mesh together when they press the #Goodweird button a few too many times and things get really weird, with male Indian line dancers in white Russian ballet leotards, American football mascots jumping in the streets of Moscow, and hairy Russian men running around in cheerleader outfits. Can't picture it? Watch the video[3]—it is definitely weird…and pretty good. At the very least, it is engaging and nearly impossible to stop watching (Fig. 12.1).

When asked about the success of the #Goodweird campaign, Roman explains results varied from market to market: "Where it was proportionately more successful were in countries that I never would have expected. In the Middle East—Pakistan specifically was a country where the #Goodweird idea really resonated. The same concept worked very well around the world and in some countries, it even got more of an amplification simply because of their proportionately higher daily use of social media like what we saw in India, Russia and Brazil."

Roman goes on to say, "Sometimes you have things that don't work the way you thought they would—little visibility, little traction. We've seen both sides. Ultimately, you can't have expectations that are too rigid. When you take a traditional campaign, you can set realistic expectations around who's going to see it, what you expect them to do—with engagement campaigns you don't have that ability. If the purpose is to build the relationship, then you have to stay true to that and not try to drive it specifically around traditional campaign parameters. #Goodweird was one of

Fig. 12.1 Goodweird campaign (*From left to right: EeOneGuy, The Viral Fever, Matthew Santoro*)

those instances where everything came together to generate a truly fantastic global result."

Incorporating Influencers into a Global Marketing Strategy

This chapter is not the first time this book highlights the emergence of influencers around the world as a growth opportunity for brands from one country to reach audiences in other countries. Chapter 4 especially argues that the rise of new technologies, coupled with the borderless nature of the Internet, makes influencer marketing a truly global phenomenon. However, the influencer landscape is developing at different rates in different markets, due to a variety of reasons related to factors such as local cultural and linguistic nuances, social media platform preferences, variations in local market size and Internet speeds, and potentially restrictive talent–agency relationships.

Or as Simon Kemp, founder of the marketing consultancy Kepios, points out: "It's increasingly easy to get broader geographic influence, simply because the internet has made connections across borders way easier, but that doesn't mean that culture is shifting at an equal pace as that connectivity." The reality is, many of the challenges that currently prevent brands from easily collaborating with influencers outside their home market will become less and less over time—both due to natural demographic shifts in the markets themselves and as result of some of the emerging technologies (especially related to overcoming language limitations) that are introduced in Chapter 13: *What's Coming—The Future of Influencer Marketing*.

Given the inevitable fall of the barriers to global participation, then what does this mean for brands today? As with anything in business, the early movers that sow seeds today will have the most to gain in a future, more globally integrated influencer marketing environment. Take to heart one of the key takeaways from Chapter 9 on collaborating with influencers—A-B-E: Always Be Experimenting—it is the only way to learn what works and what does not. By applying these learnings over an extended period of time, especially in new overseas markets, brands can identify opportunities to open up new markets for their products and services when the timing is right.

Think back to the ZooEnglish.com case study where its California-based head of China digital marketing achieved record overseas sales growth by developing relationships with the right Chinese WeChat influencers focused on their target market of early childhood education. The greatest gains will go to the first industry player to come up with the right strategy and set of tactics for a certain market. Subsequent attempts by their competitors will not lead to the same outsized returns when it comes to lead generation, sales, brand awareness, or any other goals set for international market expansion with the help of a thoughtfully designed influencer component (Fig. 12.2).

However, brands come in all shapes and sizes, from their industry to company size to relative experience/level of comfort with international business. There are many multinational brands, like Unilever or IBM, with multiple decades of experience operating in countries around the world. At the same time, there are smaller high-growth, largely domestic-focused companies with direct-to-consumer e-commerce models, or those focused on rapidly developing technologies like augmented reality that are considering expanding into their first set of international markets. Both these established and emerging types of companies have a lot to gain from considering how influencers in overseas markets could help fuel their next stage of growth internationally. But, building an extensive, globally integrated campaign the likes

"Any type of new international investment decision or resource allocation should begin first with a broader prioritization exercise as companies only have so much budget, resources and management talent to take on new initiatives ."

-Richard Leggett (@RichLeggettFSG)
Chief Executive Officer
FRONTIER STRATEGY GROUP

DigitalInfluenceBook.com   SHARE THIS

Fig. 12.2 Richard Leggett

of the Lenovo #Goodweird example can be daunting for any company to consider—influencers in multiple countries, speaking multiple languages across multiple time zones—that is a lot of complexity for a company that may not have much experience working with influencers in their home market, let alone in a foreign country.

So how can brands get started? What does it take to lay the foundation for future influencer-enabled international growth?

"Multinational companies understand their home markets best, but often experience challenges when they try to apply the same approaches, principles and operating assumptions from their home markets to international markets—especially developing markets," explains Richard Leggett, chief executive officer at Frontier Strategy Group. "The approach used to win in a developed market like the US or Europe will almost always fail to account for critical local nuances especially in diverse emerging markets across Asia, Latin America, Middle East and Africa."

Leggett's advice to executives in the process of considering piloting an international-oriented influencer relations program: "Any type of new international investment decision or resource allocation should begin first with a broader prioritization exercise as companies only have so much budget, resources and management talent to take on new initiatives especially on an international scale. It's a big world so executives need to carefully understand which local markets represent the greatest opportunity across the short-, medium- and long-term and then prioritize strategy and resource allocation accordingly."

Once a brand determines which international markets to pilot influencer marketing efforts, it needs to decide on how it intends to manage implementation. Again, the international element makes things more complex because of different cultures, communication preferences, and many of the other reasons listed earlier in this chapter, which means brands need to put in additional time to build the appropriate implementation plan. A major factor for brands to consider is how much control they want to have at the "global level" (a centrally managed approach at company headquarters) versus the "local level" (more decentralized approach, offering more decision-making authority to the local teams in the local market).

Generally, this type of planning will result in one of the following three options:

1. Local strategy with local implementation.
2. Global strategy with local implementation.
3. Global strategy with global implementation.

"I always recommend global strategy with local execution," explains Delphine Reynaud, vice president of influence at Traackr. "If you want to launch something in China, hire a local team or agency that can help localize your global strategy for that market." As the Lenovo #Goodweird example illustrates, a global framework helps ensure a consistent strategy layer is applied in all markets, and local implementation (for #Goodweird: USA, India, Russia) helps make sure people with the right understanding of the local market are able to translate the global vision in a locally relevant manner.

"I can't emphasize enough that you need to take into account the cultural difference of the people you want to engage," Reynaud says. "For example, in Latin America they're big Instagram users; so some social networks in different countries are going to be more popular than others – you're even going to have some local social networks that don't really exist outside specific markets. Do local users prefer open platforms like Instagram or closed platforms like WeChat? Getting language right is a major issue as well. You have basic trust, transparency, confidence – but the way you communicate this locally isn't going to be the same from market to market. If you don't have someone who knows the local market you're trying to get into, team up with someone who has the necessary skills – the more your strategy is clear and your objectives are clear, the easier it's going to be to find partners that can help you execute on the ground at the local level."

The key takeaway from Leggett and Reynaud's comments is that businesses big and small, B2B and B2C, all face the same challenge: There is only so much time, attention, and budget dollars to go around. Prioritization is everything. After deciding what markets to focus on, companies then need to determine how much control they want over the implementation: local–local, global–local, or global–global. Therefore, brands should set the right goals and manage expectations with internal management to gain the necessary international experience to build on over time. The path to developing an effective influencer program that incorporates influencers from overseas markets will vary from company to company.

For example, Walter Jennings at Huawei (see Chapter 6) spent about a year to build his 100-person global influencer community, whereas Yuping He from ZooEnglish.com (see Chapter 4) spent a few months before her China-focused influencers were producing sales. Unfortunately, there is not a prescriptive path that all companies can follow since each company's goals will vary substantially, and the ease of doing business in different countries varies greatly as well.

Therefore, marketers need to constantly keep their eyes open to learn what is working for others and what they might be able to apply to their own business. That is why the remainder of this chapter focuses on sharing examples of how companies have successfully worked with influencers outside their home market as part of their international marketing strategy. Review them carefully; they are filled with valuable inspiration to jump-start any company's planning for collaborating with influencers in less familiar overseas markets.

Case Studies: Cross-Border Influencer Collaboration Inspiration

An entire second book could be written to address the breadth of industries and markets affected by influencer marketing, as well as go into detail regarding how the companies involved went through the process of identifying the right influencers, engaging and collaborating with them, and measuring results. As opposed to choosing a handful of examples and going into great detail, this chapter will present a more wide-ranging set of examples. The following are 11 examples of cross-border brand-influencer collaboration to conclude Chapter 13. Just realize, they are here for inspiration, not as an exact blueprint of exactly what to do (Fig. 12.3).

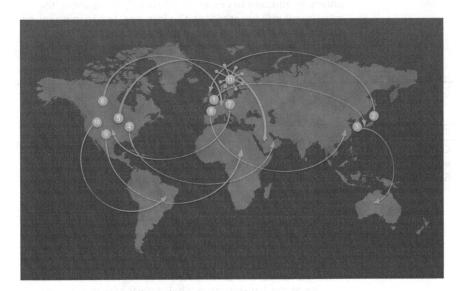

Fig. 12.3 World map case study overview

Fig. 12.4 Soukaina Aboudou's PMD Kiss Collaboration

#1 British brand -> Japan

Company: Dyson
Industry:
  Consumer tech
Goal: Product
  sales

The British consumer technology company best known for its vacuum cleaners and hand dryers continues to build an extensive YouTube influencer program in Japan. It works with a local team to engage influential consumer technology-focused influencers. In addition to being filmed entirely in Japanese, the subsequent influencer videos also incorporate engaging sound effects, animations, and other culturally relevant elements. One of the more prominent Category Influencers that Dyson includes in its influencer promotions is a highly animated consumer electronics-focused influencer named Hikakin[4]

#2 Japanese brand -> UK

Company: Uniqlo
Industry: Retail
Goal: Reputation
  management

When it relaunched its flagship store on Oxford Street, Japanese retailer Uniqlo wanted to use the opportunity to highlight the brand's deep ties to London. The original store was the first international outlet Uniqlo opened nearly 10 years earlier, and to associate its brand with the distinct London culture, Uniqlo assembled a group of six London-based Category Influencers, ranging from local artists to photographers to musicians, to serve as Cultural Ambassadors. The six influencers promoted the collaboration on their social media channels and participated in exclusive experiences held on-site at the flagship store. Their efforts helped highlight Uniqlo's long-term commitment to the city of London, and also successfully relaunched its Oxford Street location with a more modern look, and new technology-enabled shopper experiences

---

#3 Korean brand -> Australia

Company: LG
Industry:
  Consumer tech
Goal: Product
  launch

LG, the Korean smartphone and consumer electronics company, developed a campaign to launch its new LG G4 mobile phone, targeting the Australian market. It developed a six-week pilot campaign called the "G4 Recommender" program, which included 30 Australian social media influencers across different content categories (food, fashion, photography, etc.). Each week, the G4 Recommender received a new photograph challenge related to the G4 phone. LG was satisfied with the results of one of its first influencer engagements in Australia: "We were delighted with what we saw from the G4 Recommenders as an initial trial. The level of engagement and creativity of the posts exceeded our expectations. Judging by the feedback we have been able to convert customers across to our brand and created an independent voice to support awareness building for the G4," commented Brad Reed, communications marketing manager at LG Australia[5]

#4 American brand -> United Arab Emirates	
Company: PMD **Industry**: Beauty **Goal**: Product launch	PMD, a fast-growing American beauty brand that produces a line of personal beauty devices, engaged influencers to launch its new product—"PMD Kiss," an anti-aging lip treatment. Although the firm only has an official presence in the USA, UK, and Australia, it engaged influencers all around the world as part of the product launch. Soukaina Aboudou, a UAE-based beauty influencer who participated in the product launch campaign, was impressed by how "hands off" PMD was: "They only asked to review the final video, otherwise I was free to say whatever I wanted, and produce content that I knew my audience would find valuable" (Fig. 12.4)

#5 American brand -> Middle East	
Company: Pinkberry **Industry**: Franchise **Goal**: Brand awareness	Pinkberry, the American frozen yogurt franchise, launched several outlets in major metropolitan areas across a global landscape. In the USA, Pinkberry gained brand recognition from genuine "Pinkberry fan" celebrities, such as soccer talent David Beckham and renowned actress Reese Witherspoon, whose public love for Pinkberry's product created preliminary brand awareness and credibility for the franchise. Ryan Patel, who served as Pinkberry's vice president of global development at the time, says, "we adopted a similar approach across the MENA region to Peru because consumers based in these local markets traveled regularly to major cities where we already had a presence. We ended up with a mix of both prominent and local micro-influencers who shared authentic reasons why they loved our product, which then created credible online and offline word of mouth recommendations to propel our success in their regions"

#6 Canadian brand -> Europe	
Company: Lululemon **Industry**: Apparel **Goal**: Brand awareness	Lululemon, the Canadian athletic apparel company, established a Global Yoga Ambassador program comprised of seven established yoga teachers/Category Influencers with strong social followings. In order to build brand awareness in key European markets, the brand took a van full of ambassadors to six cities[6] (London, Stockholm, Munich, Copenhagen, Zurich, and Paris), where the influencers led high-profile yoga workshops. The trip became the basis of a more formal long-term-oriented influencer relations program, where the group is prominently featured on the brand's website[7] and "activated" for similar opportunities to collaborate on campaign activities when they arise

#7 German brand -> USA	
Company: SAP **Industry**: Enterprise tech **Goal**: Demand generation	SAP, the German enterprise technology company, invites US-based influencers to its annual Sapphire event in Orlando, Florida. It works with these influencers on an ongoing basis to produce content such as blog posts, webinars, and Facebook Live streaming videos to fuel demand generation across its product portfolio for various functional roles (CIO, HR, etc.). By focusing its efforts on influencers who hold the attention of SAP''s target buyers, the firm leverages its influencer relations program for ongoing prospective customer engagement to stay top-of-mind given lengthy B2B purchase cycles

#8 French brand -> China	
Company: Givenchy **Industry**: Fashion **Goal**: Product sales	The French fashion brand Givenchy collaborated with Chinese fashion Category Influencer Tao Liang, a.k.a. "Mr. Bags," to launch a limited edition (80 total) pink luxury handbag in China. Because Chinese online followers place a lot of trust on their influencers, they tend to make purchases with relative higher frequency based solely on influencer recommendations. As a result of Mr. Bags' targeted following, his posts on local platform WeChat led to the pink handbag completely selling out within only 12 minutes, generating at total 1.192 million RMB ($173,652) in sales with the bags priced at 14,900 RMB ($2170) each[8]

#9 American brand -> Africa	
Company: Intel **Industry**: Technology **Goal**: Product launch	The American technology brand prepared to launch a new line of "Intel-powered" 2-in-1 laptop/tablet devices in select markets around the world including Africa. The local Intel Africa team recruited eight social, photography, tech, music, and entertainment influencers to participate in a campaign called #InsideOut. The influencers developed original content to build awareness around the product launch with three main influencer developing their own YouTube series related to the campaign that generated approximately 150 K local views on Intel Africa's YouTube channel[9]

#10 American brand -> Latin America	
Company: Tinder **Industry**: Online app **Goal**: App downloads	The American dating and friend-finding app Tinder wanted to expand its penetration of key markets in Latin America. It hired an agency, GDW Strategies,[10] to develop a global influencer strategy focused on Latin America that involved a series of local YouTube influencers. Examples of participating influencers included Kika Nieto, a Colombia-based designer/photographer, and Kathy Castricini, a video blogger based in Rio de Janeiro, Brazil. Through collaboration with Local Influencers, Tinder was able to develop a broad selection of local YouTube video content to drive awareness of its app through tapping into the audiences of these key influencers

#11 Swedish brand -> Eight Global Markets	
Company: Absolut Vodka **Industry**: Alcohol **Goal**: Brand awareness	The Swedish vodka brand (currently owned by the French Pernod Ricard group) focused on general brand awareness in eight high-priority markets around the world: Canada, Germany, Mexico, South Africa, South Korea, UK, USA, and Brazil. It developed a brand ambassador program with Local Influencers in each of these markets, engaging them in a 17-week-long Instagram campaign called #AbsolutNights, where the influencers post creative photographs and include a caption about what made it "an Absolut Night." With the "awareness" focus, they recorded high levels of vanity metrics in each market, surpassing original targets and generating 243 posts and 340,884 interactions (likes/comments/mentions)[11]

Think Global Today and Be Prepared for Tomorrow

When David Roman and his team at Lenovo launched the #Goodweird campaign to promote its new Yoga tablet, they could not predict exactly how the results would turn out. They obviously had an initial set of assumptions, and ideal outcomes in mind, but there were a lot of moving parts—influencers based in different countries, speaking different languages and working around the clock across time zones. In the end, the Lenovo team had to put their faith in the fact that they established the right global framework upfront that could be implemented effectively by the local teams—and that is exactly what happened. Marketers should not read this case study and ask, "How can we do what Lenovo did?"

Of course, if they already have a wealth of experience building and managing influencer programs and campaigns in their home market, then something like #Goodweird may not be that unreasonable. However, most marketers should be asking themselves, "How can my company learn from what Lenovo did to pilot an influencer program that engages Local Influencers in overseas markets?" Starting with a pilot is key—working with influencers effectively at home is challenging, but throw in all of the additional complexity of foreign markets and it should be clear that a test-and-iterate or A-B-E approach is the best way to build international capacity over time.

The pilot countries or region of focus should not come about at random from a top-down management decision, but rather the natural outcome of a thoughtful prioritization process to determine which markets make the most sense to focus on in the short-, medium-, and long-term. From there management needs to determine which implementation approach they are most comfortable with: local–local, global–local, or global–global.

Remember, the 11 case studies on cross-border influencer engagement are merely meant to serve as inspiration and to help illustrate how "global" influencer marketing has and will continue to become. As the next, and final, chapter introduces—the future influencer landscape will naturally become more global, many of the current challenging aspects of collaborating with Local Influencers will dissolve, and brands will have no choice but to start looking global. The first-movers will reap outsized returns, so brands should not wait too much longer to begin—the time for businesses to prepare for this future globally integrated influencer marketing environment is today.

Marketer's Cheat Sheet

- **Start Small, Test, Repeat, Scale**: The Lenovo #Goodweird example demonstrates what is possible when a brand integrates influencers from a range of countries and cultures into their international marketing strategy in order to launch a global product. However, it also sets a high bar for companies just beginning to think about engaging influencers outside their home country. Therefore, brands should start small, apply learnings, and increase complexity over time rather than take on too much too soon.
- **Prioritize Markets Upfront**: While it may be tempting at the outset to test collaborating with influencers in a wide range of markets, brands should take a more measured approach, conducting an in-depth prioritization exercise to determine which markets are best suited to pilot. Perhaps the brand already has a local subsidiary in a given market, or maybe influencers in a market may have more shared cultural similarities (e.g., US–UK, China–Singapore) that will make collaborating and communicating much easier.
- **Align Implementation with Expectations**: When a brand is ready to implement an influencer program with international elements, its management needs to determine how much power they are willing to give to the local markets versus the global headquarters. The three choices are local–local, global–local, or global–global. Of the three, experts recommend global–local as the best way to establish a strategy at headquarters that can then be tailored for the local markets.
- **Seek Out Inspiration**: There is no textbook for "how to work with influencers" in any industry, markets of focus or at specific companies. Marketers need to constantly seek out inspiration from what other companies are doing well (like the 11 examples in this chapter) and apply relevant takeaways to build the textbook from scratch. Brands should not think so narrowly about market or industry match and should try to mine for the critical insight that could be applied in the context of their businesses.
- **Global as the New Normal**: As overwhelming as it may seem to consider working with influencers in foreign markets, global will soon be the norm, so the time to start getting accustomed to it is now. The Internet is borderless, and the cultural and linguistic barriers will disappear over time so time is of the essence for brands to look beyond their home market and gain experience collaborating with influencers overseas.

Notes

1. Slefo, George. "New Lenovo Campaign Plays Up Weird in the Good Way." *AdAge*, 30 Sept. 2015, adage.com/article/advertising/lenovo-s-latest-campaign-weird-good/300669/.
2. EEONEGUY—#ДЕЛАЙПОСВОЕМУ. *EeOneGuy*, 1 Dec. 2015, www.youtube.com/watch?v=adfO32D45Xs.
3. Three Continents, One #Goodweird Dance Off (ft. EeOneGuy, Matt Santoro and The Viral Fever). *Lenovo*, 1 Dec. 2015, https://www.youtube.com/watch?time_continue=78&v=t2JVHOmWATU.

4. ダイソンの掃除機買ってゴミ吸いまくってみた! dyson v6 fluffy+. *HikakinTV*, 27 Nov. 2015, https://www.youtube.com/watch?v=0hAVKDVXRwc.

5. LG Australia. "LG Lays Down the Challenge." *Hypetap*, 12 Sept. 2015, hypetap.com/media-releases/LG_LAYS_DOWN_THE_CHALLENGE_ WITH_G4_RECOMMENDER.pdf.

6. Karmali, Sarah. "Lululemon Launches European Yoga Tour." *Harper's BAZAAR*, 24 Aug. 2015, www.harpersbazaar.com/uk/beauty/fitness-wellbeing/news/ a35256/lululemon-launches-european-yoga-tour/.

7. Lululemon. "Global Yoga Ambassadors." *Lululemon Global Yoga Ambassador Page*, shop.lululemon.com/ambassadors/_/N-1z141e2Z1z13tru.

8. Brennan, Matthew. "1.2 Million Sold on WeChat in Just 12 minutes!" *China Channel*, 7 Feb. 2017, chinachannel.co/wechat-givenchy-influencer-case-study/.

9. Sharman, Kirsty. "South African Brands That Use Influence Marketing and Won." *BizNis Africa*, 10 Feb. 2016, www.biznisafrica.com/south-african-brands-that-use-influence-marketing-and-won-watch-video/.

10. GDW Strategies. *Tinder—Meeting People.* http://gdwstrategies.com/casestudies/.

11. Relatable. "#AbsolutNights on Instagram 2016." *Relatable*, relatable. docsend.com/view/ypuaxfy.

13

What's Coming: The Future of Influencer Marketing

Studies say people are happiest while planning a trip. Visions of tropical islands, home-cooked meals, or skiing in the mountains fill them with joy as they check ticket prices online. What they do not want to think about, however, is travel insurance. It is a topic that people usually go out of their way to avoid—while rushing through the airline ticket purchase process, most of us do not want to read through even more fine print explaining the "added benefit" of protecting our travel purchase with an insurance upsell. We hastily select "proceed without insurance" and move on.

This is the problem that Daniel Durazo, director of communications for Allianz Worldwide Partners, works on every day. His business is part of the large German global financial services company Allianz, and he is responsible for building awareness of the benefits Allianz travel insurance can bring travelers in protecting their purchases and ensuring them safe journeys.

Recognizing the power of word of mouth in the travel industry, in order to have the greatest impact on driving awareness of Allianz's travel insurance solutions, Durazo decided to turn to the voices of trusted peers, as opposed to focusing on traditional corporate, top-down marketing communications. He focused on travel bloggers, since they possess dedicated audiences of travel aficionados who turn to them for advice such as what places to travel to, which gear to buy and travel safety. Durazo and his team began researching travel bloggers online to learn more about their content, personal backgrounds, and target audiences. The result of their identification exercise led to a short list of travel influencers who appeared to be a good fit for the Allianz brand.

© The Author(s) 2018
J. Backaler, *Digital Influence*,
https://doi.org/10.1007/978-3-319-78396-3_13

As an initial pilot program, Durazo selected a relatively small number of bloggers that could grow over time depending on results. In 2015, he went live with the first version of Allianz's influencer relations program, called "The Travel Trifecta"—a group of three prominent travel bloggers: John DiScala of JohnnyJet.com, Lee Abbamonte of LeeAbbamonte.com, and Gary Arndt of Everything-Everywhere.com. Each of the three bloggers is established Category Influencers for travel, with dedicated relevant followings of travelers—Allianz's prospective customers for travel insurance (Fig. 13.1).[1]

As part of the program, the three travel influencers produced a wide range of content for Allianz including writing blog posts, hosting contests, and engaging with their fans on Facebook, Instagram, Google+, and Twitter to discuss travel insurance and its benefits, beyond trip cancellation. Johnny Jet and Lee Abbamonte led a monthly Twitter chat at #TravelHappy where they held a regular Q&A session to answer questions about travel and help foster positive connections between their audience and the Allianz brand.

In the more than three years that have passed since launching "The Travel Trifecta," Allianz's influencer relations program has expanded to include over a dozen prominent travel influencers who participate in similar activities. The participants are now considered "Brand Ambassadors" and still include Johnny Jet and Lee Abbamonte, along with a longer roster of participants.[2]

Durazo continues to push the envelope in terms of content collaboration with the participating influencers and is always on the lookout for

Fig. 13.1   Johnny Jet traveling in the Serengeti

new technologies that can help his influencers communicate Allianz's key messages more effectively to their audiences. In 2016, Allianz began providing influencers with a "Content Capsule" that could be embedded at the bottom of blog posts that mention the brand. The Content Capsule operates like a standalone mini-website that Allianz can populate with content and links to relevant resources, as well as obtain valuable analytics data.

Abbamonte expresses his appreciation for the brand relationship in a blog post following Allianz receiving an award for their influencer program: "I feel very proud to be a part of the award-winning Allianz Travel Insurance team. Our marketing and public relations initiatives have been recognized as being tops in the industry at the prestigious Adrian Awards. As many of you know, I, along with my friend and business partner, Johnny Jet, have been global ambassadors for Allianz Travel Insurance for 3+ years now. From the original Travel Trifecta, to its current form of more than a dozen diverse and creative travel bloggers and influencers, I am thrilled to be a part of this great team."

Today, Durazo manages an established influencer relations program within Allianz. Rather than a string of one-off campaign collaborations, Durazo and his influencer community are "always on," with long-term influencers like Abbamonte and Johnny Jet personally invested in the long-term success of the program. As new technologies emerge, especially ones that improve audience engagement and analytics like the Content Capsules, Allianz is in a position to adopt and adapt and strengthen the program over time.

Influence 2.0: An Always on, Technologically Enabled Future

Durazo's experience implementing an effective influencer program at Allianz helps shine a light on where the future of influencer marketing is heading—"always on," always adapting to new technologies. Allianz works with its influencers on an ongoing basis over many years; when new technologies emerge, like the Content Capsules, it adopts them and enhances the connection between its brand and the influencers' audiences. More and more technologies will continue to shift the relationship between brands and influencers, as well as influencers and their respective audiences.

As we look toward the future of influencer marketing, these are the two areas to watch: how companies will adopt the best practices introduced in this book to build ongoing influencer relations programs focused on long-term engagement, and how brands and influencers will adopt new technologies that will help improve the way both sides collaborate and ultimately engage their respective target audiences.

Influencer Relations: Long-Term Brand-Influencer Engagement

"Influence isn't a switch; it doesn't go on and off. It requires continuous care through an influencer relationship management (IRM) platform and dedicated resources to connect customers with the people who influence them every day," argues Brian Solis, principal analyst at digital research company Altimeter. He argues if brands focus too much on short-term, transactional relationships with influencers, "influence marketing succumbs to iteration rather than innovation – doing the same but with new tools versus doing new things to unlock value – without breaking new ground and introducing new value to the organization."

What he is getting at is this—by focusing on short-term, transactional campaigns, influencer marketing becomes more of an iteration of traditional celebrity endorsements, whereas if influencer marketing is implemented in a more strategic fashion, it has potential to reshape how brands connect to their audiences across all aspects of their business, from product development to employee advocacy to customer experience.

Going forward, this is why influencer relations is so important—instead of "here today, gone tomorrow" influencer campaign collaborations that dominate the focus of most influencer marketing activities to date. Thinking back to Chapter 7 and all the work that is required for brands to identify and build a relationship with the perfect influencer, once an influencer proves to be a good fit, why wouldn't a brand want to harness the long-term benefits of building an ongoing relationship with that influencer for many years, instead of watching a competitor steal them away a few months or years after collaborating with your brand? (Fig. 13.2)

As much as the release of future technology will help improve how brands identify potential influencers or how brands measure Return On Investment (ROI), successful influencer collaboration will always require a human element with person-to-person interaction. Or as Ted Rubin, chief marketing officer of Brand Innovators, calls it, a "return on relationship." This is "the value that is accrued by a person or brand due to nurturing a relationship over time. This will demonstrate that the influencer is true to the brand, and this true relationship connection will pass through to the consumer,[3]"

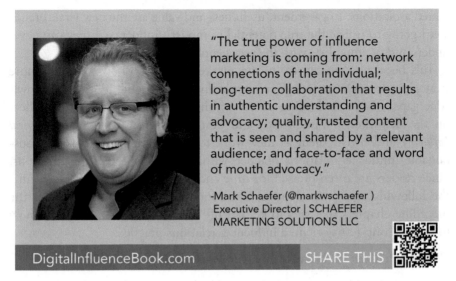

"The true power of influence marketing is coming from: network connections of the individual; long-term collaboration that results in authentic understanding and advocacy; quality, trusted content that is seen and shared by a relevant audience; and face-to-face and word of mouth advocacy."

-Mark Schaefer (@markwschaefer)
Executive Director | SCHAEFER MARKETING SOLUTIONS LLC

DigitalInfluenceBook.com   SHARE THIS

Fig. 13.2 Mark Schaefer

he says. Assuming a brand is working with the right influencers who share their target audience, brands can get much more out of that influencer than a onetime burst campaign collaboration.

Longer-term relationships are positive for influencers as well. YouTuber Taryn Southern explains that there is a better outcome when an influencer is not just paid to put out one piece of content, but is instead involved in multiple content collaborations over time: "First, it's more effective – brands know the more [often] audiences see something, the more likely they are to interact with it, and the more comfortable they get with it, and there's that name association. It's also so much better for the influencer to associate their personal brand with a company for a longer period of time. The influencer will become more comfortable with how their interactions and the conversation goes over time which will seem more authentic to their audiences and lead to better results for the brand."

This is exactly the approach to influencer engagement that Solis of Altimeter recommends in the company's study *Influence 2.0: The Future of Influencer Marketing*. The report outlines how brands can align influencer marketing efforts to the objectives of the business, its influencers and its customers, or, as he explains: "In the next iteration of [influencer] engagement, brands must think beyond "influencer marketing" to invest in real relationships with real people who are in control of their online experiences. This is no longer an attention game, but instead one of relevance, common interests and shared aspirations. Engagement, usefulness and value are the new viral. Brands aren't going to realize the true potential of modern influence if they can never understand or empathize with the very people they're hoping to reach."

This book was written to inform marketing leaders how to take a more strategic approach to influencer marketing, and to fill in critical knowledge gaps. Marketers leading efforts within their companies need to reflect on the definitions, frameworks, and best practices introduced in *Digital Influence* and revisit them as they work toward implementing a more sustainable influencer relations practice within their respective organizations. Since readers are often pressed for time and want to get "right to the point," the following summarizes the number one key takeaway from each of the preceding chapters that should be part of the internal discussion about how to build a long-term-oriented influencer relations practice:

Chapter 2: Then vs. Now: Influencer Marketing (Re-)Defined

- *Interruption Marketing vs. Influencer Marketing*: We are transitioning from an era of Interruption Marketing (where advertisers intentionally disrupt people's activities to plant a branded message into their subconscious) to one where community and word of mouth play a much more significant role (especially Influencer Marketing). Individuals do not trust brands, they are increasingly using pop-up blockers, and they want to learn from trusted peers.

Chapter 3: Levels of Influence: Key Characteristics of Modern-Day Influencers

- *Levels of Influence*: While people try to simply call everyone with an online community an "influencer," it is more informative to first break down influencers by their audience size to classify them as either Celebrity Influencer, Category Influencer (emerging or established) or Micro-Influencer. It is important to normalize audience size to account for fake followers and don't confuse general "celebrity" for actual influence.

Chapter 4: A Global Phenomenon: The Rise of Influencers Around the World

- *A Global Phenomenon*: Influencer Marketing is expanding in countries all around the globe. The rise of new technologies and borderless nature of social media platforms has fueled an international movement. While markets like the USA and UK are most established, areas like influencer-driven sales in markets like China are more advanced and show what may be possible in other markets in years to come.

Chapter 5: Business to Consumer (B2C) Influencer Marketing Landscape

- *Influencers are Brands*: Especially in a B2C context, influencers are their own personal brands; therefore, when a corporate brand wants to collaborate with an influencer, it is really two brands trying to come together, and those two brands could have very different values or ways of engaging their target audiences. Certain B2C verticals, especially toys, fashion, and beauty, are shifting budget dollars to influencer marketing in a big way, as their consumers turn to trusted online voices for guidance about what to buy and the latest trends.

Chapter 6: Business to Business (B2B) Influencer Marketing Landscape

- *B2B is as Important as B2C*: There is a common misperception that influencer marketing is only relevant for B2C firms—influencer relations is also critical for B2B brand success as they manage their corporate reputation and stay top-of-mind among potential buyers over drawn-out sales cycles. Just keep in mind, the objectivity of B2B influencers in the marketplace is the basis for everything they do. They are unable to outwardly favor one brand over another, and they have to take a much more indirect approach with regard to everything they do promote.

Chapter 7: Discover Influencers: Finding the Perfect Match

- *Target Audience First*: Many brands like to start with a "cool influencer" and then find a way to reverse engineer analysis to justify why he or she is a good fit. This approach fails every time—companies should begin influencer discovery by first defining a clear set of goals and a target audience. A large number of followers do not automatically make an influencer the right choice. Don't forget the influencer ABCC's (Authenticity, Brand Fit, Community, Content) and spend the time to research other qualitative factors including: the quality of engagement, face recognition (fame), quality of content, industry fit, and so on.

Chapter 8: Engage Influencers: Developing an Effective Outreach Strategy

- *Be Patient*: It is very easy for marketers to get excited after they identify the perfect influencer and want to reach out immediately. A generic email sent without any prior online engagement will more often than not fall to the bottom of the influencer's inbox. Without establishing clear ownership of influencer outreach, the same influencer may be contacted by multiple individuals within the same brand. This can be extremely damaging, because in the eyes of the influencer, whoever contacts them is "the brand." Companies need to make sure there are transparency and ownership around who reaches out to influencers on behalf of the brand.

Chapter 9: Working with Influencers: Potential Paths to Take

- *Freedom Within a Framework*: It is reasonable for marketers to want to control the entire creative process when collaborating with influencers. However, the entire reason brands work with influencers is that influencers know what their target audience wants. Therefore, marketers should act as directors, not as dictators, to ensure a successful collaboration. An influencer agreeing to collaborate is one thing, but engaging their audience in the right way to benefit the brand is another. Marketers need to be sure that what they ask of the influencer does not go against what their audience expects of him or her. Otherwise, if the influencer does not connect to their audience, then the whole collaboration becomes a waste of time and money.

Chapter 10: Know the Risks: The Dark Side of Influencer Collaboration

- *The Two Dark Sides*: The two dark sides of influencer marketing refer to one set of potential pitfalls related to the brand's risk exposure (such as reputational, legal and competitive risks) and one set related to the many ways supposed influencers can trick brands into working with them through faking influence (either organically through measures like pods and follow/unfollow or paying for follow acquisition and account engagement services). As governmental agencies and consumer advocacy organizations learn more about influencer marketing, the rules and regulations that brands need to abide by in different countries continue to evolve. For now, brands in most countries like the USA, the UK, and across the EU simply need to make sure their influencers use the appropriate disclosures whenever possible to ensure their collaboration cannot be perceived by outsiders as deceiving local consumers.

Chapter 11: Measure Success: What's the Return on Investment?

- *Start with Goals*: While marketers rush to report vanity metrics to justify influencer marketing spend, they need to stop and assess how the influencer activities tie back to why they started the influencer program to begin with. It all comes down to goals, and those goals will vary from company to company, and across different stages of the customer journey.

When it comes to measuring the ROI of influencer marketing, brands have tended to rely too much on vanity metrics like clicks and impressions. While these measurements have their place for broadbrand awareness efforts, most companies would be better off tying influencer activities to business outcomes like lead generation or sales (like Kimberly-Clark is doing with trackable links and coupon codes).

Chapter 12: Case Studies: Influencer Marketing Best Practices from Around the World

- *Global as the New Normal*: As overwhelming as it may seem to consider working with influencers in foreign markets, global will soon be the norm, so the time to start getting accustomed to it is now. The internet is borderless, and the cultural and linguistic barriers will decrease over time, so time is of the essence for brands to look beyond their home market and gain experience collaborating with influencers overseas. When a brand is ready to implement an influencer program with international elements, its management needs to determine how much power they are willing to give to the local markets versus the global headquarters. The three choices are local-local, global-local, or global-global—of the three, experts recommend global-local as the best way to establish a strategy at headquarters that can then be tailored for the local markets.

By keeping these points in mind while planning for future influencer relations programs, brands will be better prepared by setting more realistic expectations internally with regard to what is possible and externally by ensuring greater alignment with the influencers they seek to engage. Without these frameworks, standards and common understandings, influencer marketing practitioners will continue to promote bad habits that best-case generate a onetime burst of engagement that quickly dissipates and, worst-case, damages the brand's reputation among the influencers it fails to engage and their audiences—the very customers they want to reach.

A Look Ahead: Technology's Role in Shaping the Future of Influence

In addition to building a dedicated influencer relations practice within Allianz, Daniel Durazo also implemented new technologies that helped the influencers in his community better communicate the brand's key messages to the influencers' audiences. Going forward, new technologies will play a greater role in both resolving many of the current challenges influencer marketing practitioners face (e.g. finding the right influencer, measuring ROI) and enhancing the relationships influencers have with their audiences, in turn opening up new forms of collaboration between influencers and brands. While predictions are always tricky, the following are based on conversations with industry experts tracking these technologies and are grounded in technologies that already exist in the marketplace at the time of this book's writing.

#1 Identifying Influencers

There are currently multiple influencer databases available to identify influencers like the ones introduced in Chapter 7. However, the challenge for users is at best they establish a baseline for a few influencers to conduct more in-depth research into. Most of the current databases rely on scraping different websites for data like follower count, content relevance, and engagement rates. However, ask anyone who has used one of these databases, and they will tell you that the results vary from platform to platform. When it comes to finding influencers in less established markets (especially where local languages are preferred with minimal English-language content), there is a serious lack of relevant influencer results. This where artificial intelligence (AI) may be able to help—by developing new ways to analyze and process data across the Web to get a more complete set of influencer results, have results that take into account broader influencer content profiles instead of short-listing them based on one or two posts that mention the search query, and even assess important criteria like whether an influencer is "gaining influence" and on the rise or if they are "losing influence."

#2 International Influencers

Chapters 4 and 12 highlight the emerging opportunity for brands to work with influencers in many parts of the world as part of their international expansion strategy. However, right now it is extremely costly and complicated for brands to connect with influencers based in other countries that do not use English as their primary language. Companies either need to rely on external agencies in those markets or need to build their own Local Influencer relations team as part of their local implementation strategy. This may not be feasible for smaller companies to consider as part of their international expansion strategy, but as advances in Multi-lingual Natural Language

Processing (NLP) are applied across the platforms marketers use to engage influencers, communication across cultures will become less of an issue.

Multi-lingual NLP is a way for computers to analyze, understand, and derive meaning from human language in a meaningful way that can be applied to making translation between languages much more accurate, as marketers from one country connect with influencers in another country. Talia Baruch, chief executive officer of Yewser a firm that helps companies optimize their global growth and product performance for international markets, explains: "As multi-lingual NLP and data standardization improves, it will become significantly easier for brands to identify Local Influencers and engage with them online without needing to rely on third-party inter-mediaries." Baruch formerly led international product and growth at LinkedIn and SurveyMonkey.

#3 Reaching Out to Influencers

Chapter 8 reveals the importance of personalizing outreach communications to influencers to ensure the message comes across as uniquely written for the influencer. This generally takes a lot of upfront online research by a brand into the influencer's background and time spent reviewing their content to get a sense of how to tailor the message to elicit a response. However, there are already technologies available that will surely improve in the years to come to make this process easier.

The American technology firm Crystal has, according to its website, developed a solution: "In 2014, our team started developing applications for Crystal, our proprietary personality detection technology, in the Harvard Innovation Lab. Crystal analyzes public data to tell you how you can expect any given person to behave, how he or she wants to be spoken to, and per-haps more importantly, what you can expect your relationship to be like.[4]" Technologies like Crystal help marketers understand how to tailor their outreach communication to influencers including "words, phrases, style and tone you should use to reach the recipient in the way that they like to communicate, rather than your own – even their tolerance for sarcasm and emoticons.[5]"

#4 Working with Influencers

The ways influencers interact with their followers are limited by the current technologies available, but as they adopt new technologies like Augmented

Reality (AR) and Virtual Reality (VR), there will be new possibilities for brands to work with influencers beyond what Chapter 9 introduces. Shel Israel, chief executive officer of the Transformation Group, a San Francisco-based advisory firm that helps brands with their strategies to adopt immersive technologies like AR and VR, shares his view on how the relationship between influencers and brands is likely to evolve:

"When you look five to 10 years out, you see immersive technologies such as AR and VR evolving from expensive consumer novelties into the center of digital life. Instead of handsets, we will have headsets. Instead of the expensive, clunky things of limited capabilities we have today, we will have fashionable, powerful mobile devices that go where we go and enhance virtually all experiences.

This is murderous to traditional marketing practices. You cannot just insert a video ad into an AR stream without damaging the integrity of the experience. To be immersive, users need to be able to filter out offers, promotions and other marketing intrusions. Marketers will struggle to use new tech to get promotional messages out, to dominate the conversations, to turn customer dialogues into marketing monologs. This is what they have done in social media, where they have been fairly effective. Yet, social media has already begun this inevitable era where peer influence will dominate."

Look at the example from Chapter 12 about the Arab beauty YouTube influencer Soukaina Aboudou. Israel says, "In five to 10 years, Soukaina's followers will not have to go to a digital device, click on an app, navigate to Soukaina's video channel, and scan for a clip of her on a particular topic or talking about a company product; they will simply say her name and the digital technology will allow this influencer to digitally visit them in their living rooms. Haptic technology will allow them to hug her or just give her a friendly fist bump. She will sit and talk and interact and give beauty tips and product endorsements."

There are, of course, a great many barriers to all this. And Soukaina has a few years to wait before she makes virtual house calls throughout the Arab-speaking world, but that future is closer than many may realize. "Technology is changing so many things, but what is important here is that it is changing the nature of influence. It is changing who does the influencing, and it is remarkably elevating the power of peer influence at the expense of mass communications influence."

#5 Measuring ROI

Lastly, as Chapter 11 introduces, measuring the ROI of influencer marketing efforts is currently one of the greatest challenges marketers face. Unless they are simply chasing vanity metrics (such as likes, retweets, mentions) as part of awareness activities, measurement becomes much more difficult when trying to attribute influencer marketing efforts to business initiatives including sales, lead generation, product development, and corporate reputation management. While a subset of marketers have devised complex tracking systems that involve various coupon codes, trackable links, tracking pixels, opt-in pages all summarized in numerous spreadsheets; this approach is not sustainable—especially as brands shift to more long-term oriented, influencer relations programs. New big data analysis and interpretation technologies will help simplify the tracking process for marketers so that they can spend more time focusing on the human element of the influencer relationship while letting emerging marketing technology solutions enable more efficient measurement and ROI tracking based on the specific goals of the brand's influencer program.

Concluding Thoughts: Influence—Yesterday, Today and Tomorrow

The Allianz example that opens this chapter shows where more marketers are likely to shift their focus as they transition from an Influence 1.0 to an Influence 2.0 level of sophistication.

They are not alone, as B2C and B2B companies alike are assessing the long-term value of cultivating powerful enduring influencer relationships. Enabling technologies like those introduced in the preceding section will help to resolve many of the current challenges brands face when attempting to implement strategic, "always-on" influencer relations programs. In order to make the shift from a series of one-off campaigns that most resemble traditional celebrity endorsements, brands would be wise to adopt the best practices introduced in this book to bring a new level of standardization, implementation, and measurement to their influencer marketing efforts.

As much as modern-day influencers are pushing brands to revisit their assumptions about what it will take to engage their target audiences, influencers' significance will only increase in importance in the years to come. "Influencer marketing" is not a passing fad that can be written off as just another "new, sexy tool in the marketer's toolbox"—but rather the beginning of something much bigger. Consumers are only growing more skeptical of brands, warier of broadcast interruption marketing techniques, and as emerging technologies continue to shift where consumers spend their time and how they access information, the trusted voice of influential personalities will become one of the few remaining paths for brands to authentically grab the attention of target consumers and get their messages across.

In an increasingly digital world, digital influence is the currency that really matters. Brands need to lay the groundwork today to ensure they produce the necessary internal learnings that will generate an effective influencer program which stands the test of time and does not simply get written off as a series of tactics that didn't work out because of one or two failed attempts.

Brands must begin curating, a carefully managed, ever-expanding network of relevant influencers today, to ensure they can reach their target audiences tomorrow. Going forward, the only celebrity that matters for business is "celebrity tied to influence." It does not matter how well-known a personality is—all that matters is the idea the people you want to reach care what that person has to say about a topic that's relevant to the goals you want to achieve. If brands focus on identifying and building relationships with these types of individuals, they will hold the keys to their target consumers' attention, wallets, and ultimately trust.

Marketer's Resources

The following are a series of resources to help marketers go forth and conquer (see Appendix):

- **Influencer Software/Online Tools**: A list of some of the more widely used influencer technology solutions to identify, track, and measure influencer program activities.
- **Influencer Agencies**: A list of agencies that are both global and local that can support with various aspects of influencer marketing initiatives.
- **Image Glossary**: A full, consolidated list of all the included images and charts that appear throughout this book for your reference.

Notes

1. Allianz. "Allianz Global Assistance Announces First-of-Its-Kind Partnership with A-List Travel Bloggers, 'The Travel Trifecta.'" *Allianz Global Assistance*, 22 Jan. 2015, www.allianztravelinsurance.com/about/press/2015/travel-trifecta.htm.
2. Allianz. "Leveraging Influencers to Make Travel Protection an Easier Sell." *Allianz Global Assistance*, www.allianzworldwidepartners.com/usa/thought-leadership/leveraging-influencers-to-sell-travel-protection.

3. "An Influencer Marketing Intervention: 12 Steps to Salvaging Your Customer Relationships." *Traackr*, Apr. 2016. Web. 4 Jan. 2017, http://www.traackr.com/influencer-marketing-intervention.

4. CrystalKnows. *About Page.* https://www.crystalknows.com/about.

5. Hunt, Elle. "Crystal Knows Best … or Too Much? The Disconcerting New Email Advice Service." *The Guardian*, 18 May 2015, www.theguardian.com/media/2015/may/19/crystal-knows-best-or-too-much-the-disconcerting-new-email-advice-service.

Appendix

Influencer Software/Online Tools

The following is a selection of influencer software and online tools to consider:

BoostInsider | boostinsider.com
CrowdTap | crowdtap.com
FameBit | famebit.com
FLUVIP | fluvip.com
GroupHigh | grouphigh.com
Hypetap | hypetap.com
HYPR | hyprbrands.com
IFDB Talent Dashboard | ifdb.com
Influence.co | influence.co
Influenster | influenster.com
IZEA | izea.com
Julius | juliusworks.com
Klear | klear.com
Linquia | linquia.com
Mavrck | mavrck.co
Mention | mention.com
Neoreach | neoreach.com
Onalytica | onalytica.com
PARKLU | parklu.com
Reelio | reelio.com
Revfluence | revfluence.com
Terakeet | terakeet.com
Tinysponsor | tinysponsor.com

© The Editor(s) (if applicable) and The Author(s) 2018
J. Backaler, *Digital Influence*,
https://doi.org/10.1007/978-3-319-78396-3

Traackr | traackr.com
TRIBE | tribegroup.co

Influencer Agencies

The following is a selection of influencer agencies to consider:
360i | 360i.com
Acorn | acorninfluence.com
Clever | realclever.com
Come Round | about.comeround.com
Digital Brand Architects (DBA) | thedigitalbrandarchitects.com
Epic Signal | epicsignal.com
Evolve! | evolvesinc.com
Fanatics Media | fanaticsmedia.com
Fanology | fanologysocial.com
Goat Agency | goatagency.com
Influence Marketing Agency (IMA) | imagency.com
Ivy Worldwide | ivyworldwide.com
Mediakix | mediakix.com
Obviously | obvious.ly
Open Influence | openinfluence.com
Rare Global | rare.global
Reign Agency | reignagency.co
SocialCoaster | socialcoaster.com
Socialyte | socialyte.com
Sway Group | swaygroup.com
TheAmplify | theamplify.com
The Cirqle | thecirqle.com
TheInfluenceMarketer | theinfluencemarketer.com
Viral Nation | viralnation.com

Image Glossary

© The Editor(s) (if applicable) and The Author(s) 2018
J. Backaler, *Digital Influence*,
https://doi.org/10.1007/978-3-319-78396-3

Index

© The Editor(s) (if applicable) and The Author(s) 2018
J. Backaler, *Digital Influence*,
https://doi.org/10.1007/978-3-319-78396-3